SpringerBriefs in Philosophy

More information about this series at http://www.springer.com/series/10082

Przemysław Broniek

Computational Complexity
of Solving Equation Systems

 Springer

Przemysław Broniek
Algorithmics Research Group,
    Faculty of Mathematics and
    Computer Science
Jagiellonian University
Kraków
Poland

ISSN 2211-4548   ISSN 2211-4556   (electronic)
SpringerBriefs in Philosophy
ISBN 978-3-319-21749-9   ISBN 978-3-319-21750-5   (eBook)
DOI 10.1007/978-3-319-21750-5

Library of Congress Control Number: 2015945136

Springer Cham Heidelberg New York Dordrecht London

Printed on acid-free paper

Springer International Publishing AG Switzerland is part of Springer Science+Business Media
(www.springer.com)

Acknowledgment

I would like to thank my supervisor Prof. Paweł M. Idziak for all his advices, help and many hours spent working with me.

Contents

Abstract

We consider the computational complexity of determining whether a system of equations over a fixed algebra A has a solution. This leads to two problems, SysTermSat(A) and SysPolSat(A), in which equations are built out of terms or polynomials, respectively. We are interested in characterizing those algebras for which SysPolSat can be solved in a polynomial time. The problem has been widely studied and is open in general. All known results do not cover algebras that generate a variety admitting type 1 in the sense of Tame Congruence Theory. Since unary algebras admit only type 1, we attack the main problem by focusing on them.

We prove that the Constraint Satisfaction Problem for relational structures is polynomially equivalent to SysTermSat over unary algebras. This gives that Dichotomy Conjecture for CSP is equivalent to Dichotomy Conjecture for SysTermSat over unary algebras. Moreover, we get that the classification of SysTermSat for all algebras is as hard as the classification for unary algebras only. We also show that if $P \neq NP$ then the class of unary algebras for which SysTermSat is solvable in a polynomial time is not closed under homomorphic images. We isolate a preorder $P(A)$ to show a classification of unary algebras A with at most three elements using the width of $P(A)$. Finally we give other partial characterizations of the computational complexity of SysTermSat(A), e.g. for algebras with generic operations taking only few values. This covers a wide class of four-element unary algebras.

Chapter 1
Introduction

Abstract We introduce the reader to the problems of solving systems of equations over finite algebras and the Constraint Satisfaction Problem by providing definitions and describing the state of the art. We reference the most relevant work in the area and present existing classifications and dichotomies. We also introduce key definitions for the next chapters and give motivation for exploring unary algebras.

The problems connected with solving equations have been attracting many mathematicans for centuries. This is because equations have enough expressive power to code many problems from the nature. The search for mechanical solutions for polynomial equations occupied many people for years (Euclid, Brahmagupta, Cardano, Galois). Already in 1900 David Hilbert included, in his famous list of the most important mathematical problems, the question whether one can mechanically determine if a given diophantine equation has a solution. Matijasevič [Mat70] answered this famous 10th Hilbert problem negatively by showing that solving polynomial equations over the integer ring is actually undecidable. On the other hand the problem SATISFIABILITY, which puts foundations for the computational complexity theory, can be seen as solving equations over the two-element boolean algebra. The problem of deciding if a system of polynomial equations over fixed algebra **A** has a solution, which we call SYSPOLSAT(**A**), gains a lot of interest from computer scientists. The computational complexity of this problem has been studied and determined for several classes of algebras, but this complexity is still unknown in general. There are many results that consider particular classes of structures, e.g. two-element algebras [Sch78, GK], semilattices [JCC98], groups [GR02], monoids and semigroups [KTT07]. In all those results it is shown that SYSPOLSAT(**A**) is either in P or is NP-complete. This leads to the following:

DICHOTOMY CONJECTURE. For every finite algebra **A** the problem SYSPOLSAT(**A**) can be solved in a polynomial time or is NP-complete.

To confirm the Dichotomy Conjecture one actually needs to isolate and describe those finite algebras **A** for which SYSPOLSAT(**A**) is solvable in a polynomial time. In particular one can be interested in solving the following:

META-PROBLEM. Given on input a finite algebra **A** determine whether SYSPOLSAT(**A**) can be solved in a polynomial time.

© The Author(s) 2015

P. Broniek, *Computational Complexity of Solving Equation Systems*,
SpringerBriefs in Philosophy, DOI 10.1007/978-3-319-21750-5_1

Working in such generality, without additional assumptions on considered algebras, requires a lot of tools from universal algebra. Using such tools, Larose and Zádori [LZ06] were able to push the results for semilattices, groups and semigroups quite far. To understand what they did, and what is actually left, we need to recall some deep algebraic tools developed by Freese and McKenzie [FM87] and Hobby and McKenzie [HM88]. In particular the Tame Congruence Theory created and described in [HM88] appears to be very useful in this setting. Tame Congruence Theory is a tool for studying the local structure of finite algebras. Instead of considering the whole algebra and all of its operations at once, Tame Congruence Theory allows us to localize to small subsets on which the structure is much simpler to understand and to handle. According to this theory there are only five possible ways a finite algebra can behave locally. The local behavior must be one of the following:

1. a finite set with a group action on it,
2. a finite vector space over a finite field,
3. a two element Boolean algebra,
4. a two element lattice,
5. a two element semilattice.

Now, if from our point of view a local behavior of an algebra is 'bad' then we can often show that the algebra itself behaves 'badly'. For example, since SYSPOLSAT(**B**) for the two-element boolean algebra **B** is NP-complete then one can argue that for a finite algebra **A** that has local behavior of type **3**, the problem SYSPOLSAT(**A**) is NP-complete. On the other hand it is not true that if the local behavior of **A**, or even the entire variety HSP(**A**) generated by **A**, is 'nice' then SYSPOLSAT(**A**) is in P. Several kinds of interactions between these small sets can produce a fairly messy global behavior. Such interactions often make room for encoding NP-complete SATISFIABILITY problem. Most of the work done till now, including the one of Larose and Zádori, applies only to those finite algebras **A** for which the variety HSP(**A**) omits type **1**. In particular if HSP(**A**) omits types **1** and **5** then SYSPOLSAT(**A**) is in P if **A** is essentially a module over a finite ring and is NP-complete otherwise. The algebras that behaves like modules are called affine and are special cases of the so called (by Tame Congruence Theory) abelian algebras:

Definition An algebra **A** is *abelian* if for all its term operations t and elements $x, y, u_1, \ldots, u_n, v_1, \ldots, v_n$ we have:

$$t(x, u_1, \ldots, u_n) = t(x, v_1, \ldots, v_n) \Leftrightarrow t(y, u_1, \ldots, u_n) = t(y, v_1, \ldots, v_n).$$

However the class of abelian algebras contains also algebras that admit local behavior of type **1**. In particular all unary algebras, i.e. algebras in which all operations are unary, are abelian and in fact have only type **1** local behavior. Therefore, to solve the Meta-Problem in its full generality one needs to eventually incorporate type **1**. To do this, first one needs to understand unary algebras. As we already noticed they are abelian, actually they even generate varieties that are abelian, but, in contrast to affine subcase of abelian algebras, they admit both P and NP-complete behavior.

Indeed in Chap. 2 we will see that for unary algebras with only one operation the problem SYSPOLSAT is in P (see Lemma 2.30) while for some multi-unary algebras it is NP-complete (see Lemma 2.18). Unfortunately the situation is much worse. In Chap. 3 we actually show that confirming Dichotomy Conjecture for unary algebras is as hard as solving it in full generality. Therefore we decided to focus on special cases of unary algebras to put some light on the missing part of the full classification of the computational complexity of SYSPOLSAT.

Unary algebras have been studied by Feder et al. [FMS04] in a different context of the CONSTRAINT SATISFACTION PROBLEM. Actually solving equations is strongly connected with CSP. This in turn is an important and widely studied open problem with many practical applications, e.g. in databases. As in the case of SYSPOLSAT, and actually even older, there is the Dichotomy Conjecture for CSP over relational structures. This Dichotomy Conjecture goes back to the paper of Feder and Vardi [FV99]. Recently Klíma et al. [KTT07] proved that dichotomy for CSP reduces to dichotomy of SYSPOLSAT for semigroups. In Chap. 3 we prove a similar result for unary algebras, by showing that CSP for a finite relational structure is polynomially equivalent to SYSTERMSAT over some unary algebra. SYSTERMSAT is a slightly more general version of SYSPOLSAT involving terms instead of polynomials.

Thanks to a reduction of Larose and Zádori [LZ06] one can answer many questions for solving equations with the use of corresponding results for CSP. In particular the case $|A| = 3$ is covered by above transformation with an application of a deep Bulatov's [Bul06] result. Since this approach theoretically works in the general setting it often produces a characterization that is quite messy and hard to follow. Therefore we decided to use our own approach for three-element unary algebras. We fully characterize (in Theorem 2.26) the complexity of SYSTERMSAT(A) when A has at most three elements. This is done by isolating a preorder $P(A)$ associated with an algebra A. Our condition for computational complexity involves only the width of $P(A)$. This makes our approach more compact, as not referring to Bulatov's characterization makes it more transparent for unary algebras.

Klíma, Tesson and Thérien also provided an example of semigroups A and B such that B is a homomorphic image of A, SYSPOLSAT(A) is in P and SYSPOLSAT(B) is NP-complete. Such a phenomena exists already in the unary realm, as we show in Observation 3.7.

Finally our study on solving equations results in several partial characterization theorems for unary algebras which generic operations take only few values. In particular the case of four-element 2-valued algebras is fully solved. We hope that this brings us very close to solving the entire four-element unary algebras case.

The text is organized as follows. The rest of Chap. 1 is intended to introduce the reader to the problems of solving systems of equations and the CONSTRAINT SATISFACTION PROBLEM by providing definitions and describing the state of the art. Chapter 2 considers unary algebras and gives computational complexity characterization of SYSTERMSAT over three-element unary algebras. In Chap. 3 we present the computational complexity equivalence of CSP and SYSTERMSAT over unary algebras. Chapter 4 provides several partial characterization theorems. In Chap. 5 we present conclusions and open problems.

1.1 Definitions

Definition 1.1 An *algebra* $\mathbf{A}$ is an ordered pair (A, F), where A is a nonempty set, called the *universe* of $\mathbf{A}$, and F is a family of finitary operations over A. The operations from F are called *basic*. An algebra with finite universe is called *finite algebra*. If F is finite we say the algebra has *finite signature*.

In the rest of the text we will only consider finite algebras with finite signatures.

An operation of the form $f : A^n \to A$ is called *n-ary*. If $n = 1$ it is called *unary*. If $n = 0$ the operation f is identified with its unique value and is called a *constant*. We will denote by F_n the set of all n-ary operations from F. A more detailed description of algebraic terminology can be found in [BS81].

Definition 1.2 Let V be a set of variables. We define the set $T(V)$ of *terms* over an algebra $\mathbf{A} = (A, F)$ with variables from V to be the smallest set such that:

- $V \cup F_0 \subseteq T(V)$,
- if $t_1, \ldots, t_n \in T(V)$ and $f \in F_n$ then the string $f(t_1, \ldots, t_n) \in T(V)$.

In other words a term over $\mathbf{A}$ is a correct expression built with variables from V and operations from F. We write $t(x_1, \ldots, x_n)$ to indicate that variables occurring in t are among $x_1, \ldots, x_n$.

Terms are just syntax expressions. The following definition provides semantics of terms:

Definition 1.3 For a term $t(x_1, \ldots, x_n)$ over an algebra $\mathbf{A} = (A, F)$ we define a *term operation* $t^{\mathbf{A}} : A^n \to A$ corresponding to t by:

- if t is a variable x_i then $t^{\mathbf{A}}(a_1, \ldots, a_n) = a_i$,
- if t is of the form $f(t_1, \ldots, t_k)$ then

$$t^{\mathbf{A}}(a_1, \ldots, a_n) = f(t_1^{\mathbf{A}}(a_1, \ldots, a_n), \ldots, t_k^{\mathbf{A}}(a_1, \ldots, a_n)).$$

We define the *clone of term operations*, denoted $Clo(\mathbf{A})$, as the set of all term operations of $\mathbf{A}$. By $Clo_n(\mathbf{A})$ we denote the set of all n-ary term operations of $\mathbf{A}$.

Definition 1.4 Let V be a set of variables. We define the set $P(V)$ of *polynomials* over an algebra $\mathbf{A} = (A, F)$ with variables from V to be the smallest set such that:

- $V \cup A \subseteq P(V)$,
- if $t_1, \ldots, t_n \in P(V)$ and $f \in F_n$ then the string $f(t_1, \ldots, t_n) \in P(V)$.

Analogously a polynomial over $\mathbf{A}$ is a correct expression built with variables from V, operations from F and names for the elements of A. Since we arranged that $F_0 \subseteq A$, we have $T(V) \subseteq P(V)$, so that every term is in fact a polynomial. On the other hand we can look on a polynomial from $P(V)$ like element of $T(V)$ with some variables replaced with names for the elements of A. We can also define *polynomial operations* analogously to term operations and denote by $Pol(\mathbf{A})$ the *clone of polynomial operations* over $\mathbf{A}$. Obviously $Clo(\mathbf{A}) \subseteq Pol(\mathbf{A})$.

Definition 1.5 We define the *length* of a polynomial (term) t over an algebra $\mathbf{A} = (A, F)$, denoted $|t|$, by the following:

- if $t \in V \cup A$ then $|t| = 1$,
- if $t = f(t_1, \ldots, t_n)$ then $|t| = |t_1| + \cdots + |t_n| + 1$.

Definition 1.6 A finite *relational structure* $\mathbf{D}$ is an ordered pair $(D, \mathcal{R})$, where D is a finite nonempty set, called the *domain* of $\mathbf{D}$, and $\mathcal{R}$ is a finite nonempty family of finitary relations over D. Each $R \in \mathcal{R}$ is also assumed to be nonempty.

We will consider computational decision problems in which an algebra $\mathbf{A}$ (or the relational structure $\mathbf{D}$) is a parameter, i.e. *not* a part of the input. It makes a difference, when we consider computational complexity of such problems.

Definition 1.7 Given a set of variables V and an algebra $\mathbf{A} = (A, F)$ we define a *valuation* to be a function $v : V \to A$. With a given valuation v we define the *value of a term* for $t(x_1, \ldots, x_n)$ as $t^{\mathbf{A}}(v(x_1), \ldots, v(x_n))$. A *value of a polynomial* is defined similarly.

Definition 1.8 A *term (polynomial) equation* over $\mathbf{A} = (A, F)$ is a pair of terms (polynomials) t and s over $\mathbf{A}$, written as:

$$t(x_1, \ldots, x_n) \approx s(x_1, \ldots, x_n).$$

A *solution* of the equation is a valuation $v : \{x_1, \ldots, x_n\} \to A$ that makes the values of terms (polynomials) t and s equal. If there is such a solution we say that the equation is satisfiable.

Definition 1.9 *Problem* TERMSAT($\mathbf{A}$):
Input: Term equation $t(x_1, \ldots, x_n) \approx s(x_1, \ldots, x_n)$ over the algebra $\mathbf{A}$.
Question: Does the equation $t(x_1, \ldots, x_n) \approx s(x_1, \ldots, x_n)$ have a solution?

Definition 1.10 *Problem* SYSTERMSAT($\mathbf{A}$):
Input: A finite set of term equations:

$$S = \{t_i(x_1, \ldots, x_n) \approx s_i(x_1, \ldots, x_n)\}_i$$

over the algebra $\mathbf{A}$.
Question: Does the system S of equations have a solution?

We will also consider the following modification of the problem SYSTERMSAT($\mathbf{A}$):

Definition 1.11 *Problem* CSYSTERMSAT($\mathbf{A}$):
Input: A finite set of variables V, a finite set of term equations $S = \{t_i \approx s_i\}_i$ over V and a constraint function $C : V \to \mathcal{P}(A)$.
Question: Does S have a solution $k : V \to A$ satisfying additionally $k(v) \in C(v)$ for each $v \in V$?

By replacing term equations with polynomial equations we obtain problems POLSAT(**A**), SYSPOLSAT(**A**) and CSYSPOLSAT(**A**), respectively.

It is easy to observe that for a finite algebra **A** the problems TERMSAT(**A**), POLSAT(**A**), SYSTERMSAT(**A**) and SYSPOLSAT(**A**) belong to NP, because we can simply check whether a given valuation of variables satisfies all the equations. The main problem in this area is to characterize those finite algebras **A**, for which the problem TERMSAT(**A**) (or POLSAT(**A**), SYSTERMSAT(**A**), SYSPOLSAT(**A**)) can be solved in a polynomial time.

Since every term is a special case of a polynomial, we immediately get that for every algebra **A** the problem TERMSAT(**A**) is a subproblem of POLSAT(**A**). Thus solving TERMSAT(**A**) is not harder than POLSAT(**A**). On the other hand finding computational complexity characterization of TERMSAT is not easier than for POLSAT. Let us denote by $\mathbf{A}_A$ the algebra $\mathbf{A} = (A, F)$ extended with all constants $f \in A$. The set of polynomials over the algebra **A** is equal to the set of terms over the algebra $\mathbf{A}_A$. Thus the problem POLSAT(**A**) coincides with TERMSAT($\mathbf{A}_A$). Therefore finding computational complexity of TERMSAT($\mathbf{A}_A$) would give computational complexity of POLSAT(**A**). On the other hand not all algebras are of the form $\mathbf{A}_A$, thus one needs more effort to give computational complexity characterization for TERMSAT than for POLSAT. Therefore, surprisingly, dealing with terms gives more general results than with polynomials. The similar observations can be done for SYSTERMSAT and SYSPOLSAT.

Definition 1.12 The CONSTRAINT SATISFACTION PROBLEM over relational structure $\mathbf{D} = (D, \mathcal{R})$, denoted CSP(**D**), is a problem in which the instance consists of:

- a finite set of variables V,
- a finite set of *constraints* $\mathcal{C} = \{C_i\}_i$, each of the form:

$$C_i = (R_i, (X_1^i, \ldots, X_{n_i}^i)),$$

where $R_i \in \mathcal{R}$ and n_i is the arity of R_i. Each of X_j^i is chosen from V. The question is whether there exists a *solution*, that is a function $s : V \to D$ such that

$$(s(X_1^i), \ldots, s(X_{n_i}^i)) \in R_i,$$

for all constraints $C_i \in \mathcal{C}$.

Definition 1.13 We call a relational structure $\mathbf{D} = (D, \mathcal{R})$ *proper* if there is no element $d \in D$ such that $(d, \ldots, d) \in R$ for all $R \in \mathcal{R}$.

Consider CSP(**D**) for a relational structure $\mathbf{D} = (D, \mathcal{R})$, that is not proper, with $d \in D$ being a witness described in Defintion 1.13. Given an instance $I = (V, \mathcal{C})$ we define a valuation $s : V \to D$ by putting $s(X) = d$ for all $X \in V$. The valuation s obviously satisfies every constraint from $\mathcal{C}$ and thus is a solution of I. Because of that, for every instance of CSP(**D**) the answer for the problem is positive. Thus CSP over relational structures that are not proper can be solved in a constant time.

1.2 State of the Art

The first work on the CONSTRAINT SATISFACTION PROBLEM was definitely done by Schaefer [Sch78]. He gave full computational complexity classification of the GENERALIZED SATISFIABILITY problem, which can be seen as CSP over relational structures with two-element domain.

To present Schaefer's result we need to define several types of relations. By $\wedge, \vee, \oplus$ and $\neg$ we understand the binary operations of conjunction, disjunction, addition modulo 2 and unary operation of negation on the two-element set $\{0, 1\}$, respectively.

Definition 1.14 A relation R over two-element domain $\{0, 1\}$ is:

- 0-valid if $(0, \ldots, 0) \in R$,
- 1-valid if $(1, \ldots, 1) \in R$,
- Horn, if R is definable by a CNF formula with clauses having at most one unnegated variable,
- dual-Horn, if R is definable by a CNF formula with clauses having at most one negated variable,
- affine, if R can be defined by a system of linear equations over the two-element field,
- bijunctive, if R can be defined by a CNF formula with clauses having at most two literals.

We say that a relational structure $(\{0, 1\}, \mathcal{R})$ is 0-valid, 1-valid, Horn, dual-Horn, affine or bijunctive if all of its relations are 0-valid, 1-valid, Horn, dual-Horn, affine or bijunctive, respectively.

It is known (see e.g. [Sch78, DP92, KK01]) that:

Fact 1.15 A relation R over $\{0, 1\}$ is:

- Horn if and only if it is closed under the component-wise operation $x \wedge y$,
- dual-Horn if and only if it is closed under the component-wise operation $x \vee y$,
- affine if and only if it is closed under the component-wise ternary operation $x \oplus y \oplus z$,
- bijunctive if and only if it is closed under the component-wise ternary majority operation $(x \vee y) \wedge (y \vee z) \wedge (x \vee z)$.

Below we present Schaefer characterization:

Theorem 1.16 (Schaefer [Sch78]) *Let* $\mathbf{D} = (\{0, 1\}, \mathcal{R})$ *be a relational structure over two-element domain. If* $\mathcal{R}$ *is 0-valid, 1-valid, Horn, dual-Horn, affine or bijunctive then* CSP($\mathbf{D}$) *is in* P. *Otherwise* CSP($\mathbf{D}$) *is* NP-complete.

During the first 20 years after Schaefer's result our knowledge about computational complexity of CSP has been growing rather slowly. In the last 10 years essential progress has been made, namely CSP has been solved for many special cases. Every

solved case so far happens to give only two kinds of computational complexities, i.e.
P or NP-complete. This motivated Feder and Vardi [FV99] to state the following
Dichotomy Conjecture:

Conjecture 1.17 *For a given relational structure* **D** *the problem* CSP(**D**) *is in* P *or
is* NP-complete.

During consecutive work over CSP the connection with universal algebra, first
uncovered by Jeavons [Jea98], showed up to be very fruitfull. The main notion that
connects CSP with universal algebra is described in the following definition:

Definition 1.18 Let $R \subseteq D^n$ and $f : D^k \to D$. We say f is a polymorphism of R
(or R is invariant under f) if:

$$\left.\begin{array}{c} (x_1^1, \ldots, x_n^1) \in R \\ \vdots \\ (x_1^k, \ldots, x_n^k) \in R \end{array}\right\} \Rightarrow (f(x_1^1, \ldots, x_1^k), \ldots, f(x_n^1, \ldots, x_n^k)) \in R.$$

We say f is a polymorphism of the family $\mathcal{R}$ if f is a polymorphism of every relation
from $\mathcal{R}$. We denote by $Pol(\mathcal{R})$ the set of all polymorphisms of $\mathcal{R}$.

In other words, f is a polymorphism of $\mathcal{R}$ if f is a homomorphism of the product
relational structure $(D, \mathcal{R})^k$ into $(D, \mathcal{R})$. This can be expressed also that each n-ary
relation from $\mathcal{R}$ is a subalgebra of the power algebra $(D, f)^n$.

For a better understanding of the polymorphism concept observe that the functions
$x \wedge y, x \vee y, x \oplus y \oplus z$ and $(x \vee y) \wedge (y \vee z) \wedge (x \vee z)$ from Fact 1.15 are polymorphisms
for relations that are Horn, dual-Horn, affine or bijunctive, respectively.

The concept of polymorphism appeared to be very fruitfull when applied to CSP.
In fact all relational structures **D** for which CSP(**D**) is known to be solvable in a
polynomial time admit some non-trivial polymorphisms. On the other hand NP-
complete cases admit only essentially unary polymorphisms which are bijective.

A non-trivial algebra **A** is called a G-set if all its term operations are essentially
unary and bijective, so that they generate some group G. Switching from relational
structure **D** to the corresponding algebra $(D, Pol(\mathcal{R}))$ Bulatov et al. [BKJ00] proved
the following:

Theorem 1.19 *Let* **D** $= (D, \mathcal{R})$ *be a relational structure such that there exists a
factor-algebra of a subalgebra of* $(D, Pol(\mathcal{R}))$ *that is a G-set. Then* CSP(**D**) *is*
NP-complete.

Theorem 1.19 actually covers all known NP-complete cases of CSP. This encour-
aged Bulatov et al. [BKJ00] to reformulate the Dichotomy Conjecture by explicitly
stating the splitting condition:

Conjecture 1.20 *Let* **D** $= (D, \mathcal{R})$ *be a relational structure. The problem* CSP(**D**)
is NP-complete *if there exists a factor-algebra of a subalgebra of* $(D, Pol(\mathcal{R}))$ *that
is a G-set. Otherwise* CSP(**D**) *is in* P.

Again, after switching from **D** to $(D, Pol(\mathcal{R}))$ the tools of Tame Congruence Theory allowed Bulatov [Bul06] to confirm Conjecture 1.20 (and therefore the Dichotomy Conjecture) for relational structures over three-element domain. The complications and number of cases that have to be considered already in three-element realm show that new techniques are needed for designing polynomial time algorithms for **D**'s that are not producing G-set situation described in Theorem 1.19. One such approach was successfully developed by Idziak et al. [IMM+07].

Our brief description of CSP is motivated by the fact that solving systems of equations is strongly connected with CSP. In the Introduction we have already mentioned the following Dichotomy Conjecture:

Conjecture 1.21 *For a finite algebra* **A** *the problem* SYSPOLSAT(**A**) *is in* P *or is* NP-complete.

The first significant work in this area was done by Goldmann and Russell [GR02] for groups:

Theorem 1.22 *For a finite group* **A** *the problem* SYSPOLSAT(**A**) *is in* P *if* **A** *is abelian and it is* NP-complete *otherwise.*

Many other classes of algebras were considered:

Theorem 1.23 (Jeavons et al. [JCC98]) *For a finite semilattice* **A** *the problem* SYSPOLSAT(**A**) *is in* P.

Theorem 1.24 (Klíma et al. [KTT07]) *Let* **A** *be a finite monoid. The problem* SYSPOLSAT(**A**) *is in* P *if* **A** *is commutative and is the union of its subgroups and it is* NP-complete *otherwise.*

One of the possible approaches to attack Conjecture 1.21 was undertaken by Larose and Zádori [LZ06]. They transformed the problem SYSPOLSAT to CSP in order to use results that has been already known for CSP. Their reduction is simple enough to be presented here.

Definition 1.25 Let **A** be an algebra. For its n-ary operation f we denote by f° the *graph* of f, defined as follows:

$$f^\circ = \{(x_1, \ldots, x_n, y) : f(x_1, \ldots, x_n) = y\}.$$

If c is a constant then we put $c^\circ = \{c\}$.

Theorem 1.26 (Larose and Zádori [LZ06]) *Let* $\mathbf{A} = (A, F)$ *be an algebra. The problem* SYSPOLSAT(**A**) *is equivalent via logspace Turing reductions to the problem* CSP(A, T), *where* T *consists of all relations of the form* f°, *with* $f \in F \cup A \cup \{id\}$.

Proof Denote $\widehat{\mathbf{A}} = (A, T)$ and take an instance I of $\mathrm{CSP}(\widehat{\mathbf{A}})$ with variables $V = \{x_1, \ldots, x_n\}$ to produce an instance $I^\star$ of $\mathrm{SysPolSat}(\mathbf{A})$. For each constraint of the form $(f^\circ, (x_{i_1}, \ldots, x_{i_{n+1}}))$ we create an equation $f(x_{i_1}, \ldots, x_{i_n}) \approx x_{i_{n+1}}$. If f is the identity function then we get $x_{i_1} \approx x_{i_2}$ and if f is a constant c then we get $x_{i_1} \approx c$. Obviously the instance $I^\star$ has a solution if and only if the instance I has the solution. The construction can be done in a constant space.

Let J be an instance of $\mathrm{SysPolSat}(\mathbf{A})$. We enumerate all variables from J by $x_1, \ldots, x_n$ and start the construction of the instance $J^\star$ of $\mathrm{CSP}(\widehat{\mathbf{A}})$ by putting all the variables $x_1, \ldots, x_n$ to $J^\star$. Next, for each equation of the form $t \approx s$ from J we create new variables $y_1^t, \ldots, y_{|t|}^t$ and $y_1^s, \ldots, y_{|s|}^s$. Then we rewrite the polynomial t (and s respectively) as a sequence of constraints of the form:

$$(f^\circ, (y_{i_1}^t, \ldots, y_{i_k}^t, y_{i_{k+1}}^t)) \text{ or } (id^\circ, (x_{i_1'}, y_{i_2'}^t)),$$

where f is a k-ary symbol from T occurring in t (or s). Finally, we add a constraint $(id^\circ, (y_1^t, y_1^s))$ to $J^\star$. For instance, if the polynomials are $t = f(g(x_1, x_2), c, x_1, h(x_4))$ and $s = x_1$ then we construct the following constraints:

$$(f^\circ, (y_2^t, y_5^t, y_6^t, y_7^t, y_1^t)),$$
$$(g^\circ, (y_3^t, y_4^t, y_2^t)),$$
$$(id^\circ, (x_1, y_3^t)),$$
$$(id^\circ, (x_2, y_4^t)),$$
$$(c^\circ, (y_5^t)),$$
$$(id^\circ, (x_1, y_6^t)),$$
$$(h^\circ, (y_8^t, y_7^t)),$$
$$(id^\circ, (x_4, y_8^t)),$$
$$(id^\circ, (x_1, y_1^s)),$$
$$(id^\circ, (y_1^t, y_1^s)).$$

As previously the equivalence of the instances J and $J^\star$ is obvious. The construction of $J^\star$ can be done with use of counters that keep track of the operation symbol f being treated, the depth of the nesting, as we search for the arguments for f° and the position in the term. Thus it can be done in a logarithmic space.   □

With the help of Theorem 1.26, Larose and Zádori were able to characterize more classes of algebras with respect to computational complexity of $\mathrm{SysPolSat}$:

Theorem 1.27 *For a finite non-trivial lattice* $\mathbf{A}$ *the problem* $\mathrm{SysPolSat}(\mathbf{A})$ *is* NP-complete.

Theorem 1.28 *For a finite ring* $\mathbf{A}$ *the problem* $\mathrm{SysPolSat}(\mathbf{A})$ *is in* P *if* $\mathbf{A}$ *is a zero ring (i.e.* $x \cdot y = 0$ *for all* $x, y \in A$) *and it is* NP-complete *otherwise.*

Actually the easy transformation of Theorem 1.26 has been proved by Larose and Zádori to be strong enough to get universal algebraic generalizations of many earlier results, including semilattices, groups, lattices and rings:

Theorem 1.29 *If* **A** *be a non-trivial algebra in a congruence-distributive variety. Then* SYSPOLSAT(**A**) *is* NP-complete.

Theorem 1.30 *Let* **A** *be a finite algebra in a congruence-modular variety. Then the problem* SYSPOLSAT(**A**) *is in* P *if* **A** *is polynomially equivalent to a module, and it is* NP-complete *otherwise.*

Actually, as it has been already mentioned in the Introduction, Larose and Zádori extended Theorem 1.30 to the following one:

Theorem 1.31 *Let* **A** *be a finite algebra that omits type* **5** *and such that all algebras in* HSP(**A**) *omit type* **1**. *Then the problem* SYSPOLSAT(**A**) *is in* P *if* **A** *is polynomially equivalent to a module, and it is* NP-complete *otherwise.*

As we have discussed in the Introduction, Theorem 1.31 tells us that for affine algebras the problem SYSPOLSAT can be solved in a polynomial time. On the other hand, affine algebras form a special case of abelian algebras. Moreover, one can see that all above results do not cover algebras **A** that admit type **1** in the variety HSP(**A**). Actually, we do not know much about algebras **A** even if all algebras in HSP(**A**) admit only type **1**. Therefore our goal in the next chapter is to focus on unary algebras **A**, for which we know that they are abelian and have only type **1** in the variety HSP(**A**). We hope that the knowledge about them would put more light on the classification of the whole type **1** setting.

References

[BKJ00] Bulatov AA, Krokhin AA, Jeavons P, (2000) Constraint satisfaction problems and finite algebras. Automata, languages and programming (Geneva, (2000) Lecture notes in computer science, vol 1853. Springer, Berlin, pp 272–282

 [BS81] Burris S, Sankappanavar HP (1981) A course in universal algebra. Graduate texts in mathematics, vol 78. Springer, New York

[Bul06] Bulatov AA (2006) A dichotomy theorem for constraint satisfaction problems on a 3-element set. J ACM 53(1):66–120 (electronic)

[DP92] Dechter R, Pearl J (1992) Structure identification in relational data. Artif Intell 58(1–3):237–270

[FM87] Freese R, McKenzie R (1987) Commutator theory for congruence modular varieties. London mathematical society lecture note series, vol 125. Cambridge University Press, Cambridge

[FMS04] Feder T, Madelaine F, Stewart IA (2004) Dichotomies for classes of homomorphism problems involving unary functions. Theor Comput Sci 314(1–2):1–43

[FV99] Feder T, Vardi MY (1999) The computational structure of monotone monadic SNP and constraint satisfaction: a study through Datalog and group theory. SIAM J Comput 28(1):57–104 (electronic)

 [GK] Gorazd T, Krzaczkowski J Solving equations over two-element algebras. Submitted to Reports on Mathematical Logic. <!– Missing/Wrong Year –>

[GR02] Goldmann M, Russell A (2002) The complexity of solving equations over finite groups. Inf Comput 178(1):253–262

[HM88]   Hobby D, McKenzie R (1988) The structure of finite algebras. Contemporary mathe-
            matics, vol 76. American Mathematical Society, Providence
[IMM+07] Idziak PM, Markovic P, McKenzie R, Valeriote M, Willard R (2007) Tractability and
            learnability arising from algebras with few subpowers. LICS
[JCC98]   Jeavons P, Cohen D, Cooper MC (1998) Constraints, consistency and closure. Artif
            Intell 101(1–2):251–265
[Jea98]   Jeavons P (1998) On the algebraic structure of combinatorial problems. Theor Comput
            Sci 200(1–2):185–204
[KK01]   Kirousis, LM, Kolaitis PG (2001) The complexity of minimal satisfiability problems.
            STACS 2001 (Dresden). Lecture notes in computer science, vol 2010. Springer, Berlin,
            pp 407–418
[KTT07]   Klíma O, Tesson P, Thérien D (2007) Dichotomies in the complexity of solving systems
            of equations over finite semigroups. Theory Comput Syst 40(3):263–297
[LZ06]   Larose B, Zádori L (2006) Taylor terms, constraint satisfaction and the complexity of
            polynomial equations over finite algebras. Int J Algebra Comput 16(3):563–581
[Mat70]   Matijasevič JV (1970) The Diophantineness of enumerable sets. Doklady Akademii
            Nauk SSSR 191:279–282
[Sch78]   Schaefer TJ (1978) The complexity of satisfiability problems. In: Conference record
            of the tenth annual ACM symposium on theory of computing. ACM, San Diego, New
            York, pp 216–226

Chapter 2
Unary Algebras

Abstract We introduce the definition of unary algebra as well as subclasses of it called k-valued, strongly-k-valued and strongly-k-generated. Then we proceed with the simplification algorithm that transforms each system of equations into a more regular one at the expense of adding some definable constraints. Finally we give computational complexity characterization of SysTermSat over three-element unary algebras that depends on width of a special preorder constructed from given algebra.

Definition 2.1 We call an algebra **A** *unary* if all of its basic operations are unary or constants.

Example 2.2 The algebra $\mathbf{A} = (\{0, 1, 2\}, f, g, h)$ with the following operations is unary.

x	f	g	h
0	1	0	0
1	0	1	0
2	0	0	1

We have announced in Sect. 1.1 that considering terms, instead of polynomials, is a more general approach. Since one can switch between terms and polynomials, when necessary, we are going to use terms during the presentation in the rest of the text.

The definition of $T(V)$ over **A** is much simpler, when applied to unary algebras. Any term over a unary algebra $\mathbf{A} = (A, F)$:

- is either a variable x from V,
- or has the form $f_1(f_2(\ldots f_n(x)\ldots))$ for some variable x from V and some sequence $(f_1, \ldots, f_n)$ of basic operations from F_1,
- or is an element c from F_0,
- or has the form $f_1(f_2(\ldots f_n(c)\ldots))$ for some element c from F_0 and some sequence $(f_1, \ldots, f_n)$ of basic operations from F_1.

Since we will be solving term equations, we are interested how values of terms depend on valuation of variables. However, value of a term is defined by its term operation, thus we are going to switch from terms to term operations. A term over

P. Broniek, *Computational Complexity of Solving Equation Systems*,
SpringerBriefs in Philosophy, DOI 10.1007/978-3-319-21750-5_2

A has one or no variable, but for convenience we will treat all of them as unary term operations from $Clo_1(\mathbf{A})$.

Let t be a term over **A**. If t has one variable x then we replace t by $t^{\mathbf{A}}(x)$, where $t^{\mathbf{A}} \in Clo_1(\mathbf{A})$. If t has no variable then we need a small discussion about replacement that we make. The value of t is constant, say $c \in A$. However, we can treat t as a term $t(x)$ with some variable x. This trick gives us that unary term operation that takes only value c belongs to $Clo_1(\mathbf{A})$. Since we are going to use only operations from $Clo_1(\mathbf{A})$, we denote by c the above unary term operation and call it a constant. Then we finally replace t with $c \in Clo_1(\mathbf{A})$, but without specifying a variable.

Given a term $t(x)$ we can compute its term operation $t^{\mathbf{A}}(x)$ in time $O(|t|\,|A|)$. Every unary term operation can be described by its table of size $O(|A|)$. To compute $t^{\mathbf{A}}$ we start with the identity operation and compose basic operations that form sequence $(f_1, \ldots, f_n)$ in t. Dealing with t without a variable is even simpler.

We are going to show, that in problems that have terms on their input we can replace them by unary term operations without essential change of the complexity. If F_1 is nonempty then the number of terms over **A** is infinite. However, if we consider unary term operations, the situation is different. Every unary term operation is a unary function over A, thus $Clo_1(\mathbf{A})$ is finite. This happens because many terms define the same term operations.

We compute the set $Clo_1(\mathbf{A})$ in time $O(|A|^{2|A|})$. Each unary term operation $p(x)$ is generated by a term over **A**. Consider a shortest one, say $t_p(x) = f_1(f_2(\ldots f_n(x)\ldots))$. As $t_p(x)$ is the shortest, terms of the form $f_i(\ldots f_n(x)\ldots)$, for $i \in \{1 \ldots n\}$ generate different term operations. As there are at most $|A|^{|A|}$ term operations, $n \leqslant |A|^{|A|}$. We generate all such sequences of length at most $|A|^{|A|}$ and find $t_p(x)$ for given $p(x)$. Similarly we check all sequences of the form $f_1(f_2(\ldots f_n(c)\ldots))$, for $c \in F_0$ and $n \leqslant |A|$. As we never consider **A** as a part of the input, the complexity of the above process is constant for us. A careful implementation of the idea described above gives algorithm working in $O(|A|^{2|A|})$ time. Observe that $Pol_1(\mathbf{A})$ is $Clo_1(\mathbf{A})$ expanded by all missing constant operations. On the other hand $Clo_1(\mathbf{A})$ does not need to have all constant operations.

Example 2.3 For the algebra from the Example 2.2 the clone of unary term operations $Clo_1(\mathbf{A})$ contains 9 elements. The identity operation, corresponding to a variable, is denoted by id.

x	id	f	g	h	hh	fhh	ff	fg	fh
0	0	1	0	0	0	1	0	1	1
1	1	0	1	0	0	1	1	0	1
2	2	0	0	1	0	1	1	1	0

Observe that $Clo_1(\mathbf{A})$ contains two constant operations, hh and fhh. The clone of polynomial operations $Pol_1(\mathbf{A})$ containts also missing third constant operation equal to 2. The names of operations are not unique, as for example $hh = hf$ but it does not matter to us. For each particular element of $Clo_1(\mathbf{A})$ we can assign new symbol or choose unique term string, that defines it.

Lemma 2.4 *For a fixed unary algebra* **A** *the problems of the form* TERMSAT(**A**), SYSTERMSAT(**A**), CSYSTERMSAT(**A**) *are polynomially equivalent[1] to their counterparts in which on input the terms are replaced by term operations. The same happens with* POLSAT(**A**), SYSPOLSAT(**A**) *and* CSYSPOLSAT(**A**).

Proof Since **A** is not a part of the input there is a transformation working in linear time. Simply note that passing from a term t to $t^{\mathbf{A}}$ takes $O(|t|)$ time and passing from a term operation p to t_p takes constant time $O(|A|^{2|A|})$. $\square$

Lemma 2.4 allows us to switch between input representation by terms and term operations in each of the problem we consider. Therefore, when dealing with terms over unary algebras we will often denote by t both a term and its corresponding term operation. The same applies to polynomials and polynomial operations.

In view of all the remarks above, during the preprocessing of term equations given on input, we replace terms by their term operations:

$$x \to id(x),$$

$$f_1(f_2(\ldots f_n(x)\ldots)) \to (f_1^{\mathbf{A}} \ldots f_n^{\mathbf{A}})(x),$$

where now $(f_1^{\mathbf{A}} \ldots f_n^{\mathbf{A}})$ is an element of $Clo_1(\mathbf{A})$ and therefore has length 1. Similarly for a term without a variable we also remove unnecessary compositions and replace:

$$f_1(f_2(\ldots f_n(c)\ldots)) \to (f_1^{\mathbf{A}} \ldots f_n^{\mathbf{A}})(c) \in Clo_1(\mathbf{A}).$$

Summarizing, every member of $T(V)$ over unary algebra **A** can be replaced by $t(x)$, where $x \in V$ and $t \in Clo_1(\mathbf{A})$ or by a constant $c \in Clo_1(\mathbf{A})$. Remember all constants are expressible by polynomials, but not necessarily all of them by terms.

Definition 2.5 Given a unary term operation $t(x)$ we define: $\mathrm{Var}(t(x)) = \{x\}$, while given a constant unary term operation c with unspecified variable, we put $\mathrm{Var}(c) = \emptyset$. Moreover, for an equation $t \approx s$ we put $\mathrm{Var}(t \approx s) = \mathrm{Var}(t) \cup \mathrm{Var}(s)$. Finally, for a set of equations S, we put $\mathrm{Var}(S) = \bigcup_{e \in S} \mathrm{Var}(e)$.

After all the preparations we have just made, we are ready for the following description of possible shapes of term equations.

Observation 2.6 With respect to left-right symmetry, each term equation has one of the following forms:

- no-variable equation

  - the equation has the form $c_1 \approx c_2$ for two constants $c_1, c_2 \in Clo_1(\mathbf{A})$,

- one-sided one-variable equation

---

[1] We use here (and in the rest of the text) polynomial-time many-one reductions also known as polynomial transformations.

– the equation has the form $f(x) \approx c$ for some $x \in V$, $f \in Clo_1(\mathbf{A})$ and constant $c \in Clo_1(\mathbf{A})$,

- two-sided one-variable equation

  – the equation has the form $f(x) \approx g(x)$ for some $x \in V$ and $f, g \in Clo_1(\mathbf{A})$,

- two-sided two-variables equation

  – the equation has the form $f(x) \approx g(y)$ for some $x, y \in V$, $x \neq y$ and $f, g \in Clo_1(\mathbf{A})$.

At this moment we can easily see that solving systems of equations over unary algebra is interesting only if the number of equations is unbounded. Since every equation has at most two variables, we can solve every single equation in a quadratic time of the size of an algebra. Even solving k equations, where k is not part of the input, is polynomial in size of $|A|$. Therefore in the rest of the text we will only consider systems of equations.

Let I_1, I_2 be instances of either of the problems SYSTERMSAT($\mathbf{A}$), CSYSTERM-SAT($\mathbf{A}$), SYSPOLSAT($\mathbf{A}$) and CSYSPOLSAT($\mathbf{A}$). We say I_1 and I_2 are *equivalent* if I_1 has a solution if and only if I_2 does. Since the size of all equations is uniformly bounded, we define the *size* of an instance I, denoted $|I|$, as the number of equations.

For a finite set A and a permutation $p : A \to A$ there is k such that $p^k(a) = a$ for all $a \in A$, indeed e.g. $k = |A|!$ works. Now if $p \in Clo_1(\mathbf{A})$, where $\mathbf{A}$ is an algebra, p^{k-1} is a unary term operation and obviously p^{k-1} is the inverse of p. Thus $p \in Clo_1(\mathbf{A})$ implies $p^{-1} \in Clo_1(\mathbf{A})$.

Definition 2.7 A unary term operation t over a unary algebra $\mathbf{A} = (A, F)$ is:

- *k-valued*, if $|t(A)| \leqslant k$,
- *generic*, if $1 < |t(A)| < |A|$, so that t is neither a constant nor a permutation.

Definition 2.8 A set T of unary term operations is *k-valued* if every generic term operation in T is k-valued.

Definition 2.9 A set T of unary term operations over a unary algebra $\mathbf{A} = (A, F)$ is *strongly-k-valued* if there is $D \subseteq A$ with $|D| \leqslant k$ such that $t(A) \subseteq D$ for every generic $t \in T$.

Definition 2.10 Let (A, F) be a unary algebra. A set $C \subseteq Clo_1(A, F)$ is *strongly-k-generated* if $C = Clo_1(A, G)$ for some *strongly-k-valued* set $G \subseteq C$.

Definition 2.11 A unary algebra $\mathbf{A}$ is:

- *k-valued* if $Clo_1(\mathbf{A})$ is *k-valued*.
- *strongly-k-valued* if $Clo_1(\mathbf{A})$ is *strongly-k-valued*.
- *strongly-k-generated* if $Clo_1(\mathbf{A})$ is *strongly-k-generated*.

Observation 2.12 Let $\mathbf{A} = (A, F)$ be a unary algebra. The following diagram shows straightforward dependencies between properties defined in Definitions 2.8–2.11.

$$F \text{ is strongly-}k\text{-valued} \qquad A \text{ is strongly-}k\text{-valued}$$
$$\searrow \qquad\qquad\qquad \downarrow$$
$$A \text{ is strongly-}k\text{-generated}$$
$$\downarrow$$
$$F \text{ is }k\text{-valued} \quad \longleftrightarrow \quad A \text{ is }k\text{-valued}$$

With the next two examples we see that the implications in the right column can not be reversed.

Example 2.13 Let the algebra $A = (\{0, 1, 2, 3\}, F)$, with $F = \{p, f\}$ be defined by:

x	p	f	pf
0	1	1	0
1	0	1	0
2	2	2	2
3	3	2	2

Then the set F is strongly-2-valued. Therefore the set $Clo_1(A) = \{id, p, f, pf\}$ is strongly-2-generated. Thus A is strongly-2-generated but not strongly-2-valued, as $|f(A) \cup pf(A)| = 3$.

Example 2.14 Let the algebra $A = (\{0, 1, 2, 3\}, F)$, with $F = \{f, g\}$ be defined by:

x	f	g	fg
0	0	0	0
1	0	0	0
2	0	2	0
3	3	0	0

Then the set F is 2-valued and $Clo_1(A) = \{id, f, g, fg\}$ is 2-valued. Thus A is 2-valued but not strongly-2-generated.

2.1 Simplification Algorithm

Definition 2.15 Let $A = (A, F)$ be a unary algebra. The set of *definable constraints* over algebra A is the smallest set $\mathcal{C}(A) \subseteq \mathcal{P}(A)$ such that:

- if $t, s \in Clo_1(A)$ then $\{a \in A : t(a) = s(a)\} \in \mathcal{C}(A)$,
- if $C_1, C_2 \in \mathcal{C}(A)$ then $C_1 \cap C_2 \in \mathcal{C}(A)$,
- if $C \in \mathcal{C}(A)$ and $t \in Clo_1(A)$ then $t(C), t^{-1}(C) \in \mathcal{C}(A)$.

Note that, for a unary algebra A the set $\mathcal{C}(A)$ is computable in $O(|A|^{2|A|})$ time, because there are at most $|A|^{2|A|}$ possibilities in each case of Definition 2.15.

Our next Lemma shows that each definable constraint is essentially expressible by a set of equations.

Lemma 2.16 *Let* $\mathbf{A} = (A, F)$ *be a unary algebra,* $C \in \mathcal{C}(\mathbf{A})$ *be a definable constraint and* x *be a variable. Then there is a set* $S_C(x)$ *of term equations over* $\mathbf{A}$*, with* $V = \mathrm{Var}(S_C(x))$ *and* $x \in V$ *such that:*

- *if* $v : V \to A$ *is a solution of* $S_C(x)$ *then* $v(x) \in C$,
- *if* $a \in C$ *then there exists a solution* $v : V \to A$ *of* $S_C(x)$ *such that* $v(x) = a$.

Proof We induct on the complexity of how C is built according to the definition of $\mathcal{C}(\mathbf{A})$. If $C = \{a \in A : t(a) = s(a)\}$ then the set $S_C(x) = \{t(x) \approx s(x)\}$ satisfies the Lemma.

If $C = C_1 \cap C_2$ then first we rename all variables in $\mathrm{Var}(S_{C_2}(x)) \setminus \{x\}$ so that $\mathrm{Var}(S_{C_1}(x)) \cap \mathrm{Var}(S_{C_2}(x)) = \{x\}$ and then we put $S_C(x) = S_{C_1}(x) \cup S_{C_2}(x)$. The required properties of solutions of $S_C(x)$ are obvious.

Finally, let $C = t^{-1}(s(C'))$ for $t, s \in Clo_1(\mathbf{A})$. Since $id \in Clo_1(\mathbf{A})$ this covers the last possibility of building C. First note that the set C' can be realized by a set $S_{C'}(y)$ of equations in which x does not appear. Now put $S_C(x) = S_{C'}(y) \cup \{t(x) \approx s(y)\}$. For short put $V' = \mathrm{Var}(S_{C'}(y))$ and $V = \mathrm{Var}(S_C(x))$ and note that $V = V' \cup \{x\}$. First suppose that $v : V \to A$ is a solution of $S_C(x)$. Then $v|_{V'}$ is a solution of $S_{C'}(y)$ and by the induction hypothesis we have $v(y) \in C'$. Since v satisfies also the new equation we get $t(v(x)) = s(v(y))$, which implies $v(x) \in t^{-1}(s(C')) = C$. Now assume $a \in t^{-1}(s(C'))$ so that $t(a) = s(a')$ for some $a' \in C'$. Again by the induction hypothesis we get a solution $v' : V' \to A$ of $S_{C'}(y)$ such that $v'(y) = a'$. Then the valuation $v : V \to A$ defined by:

$$v(z) = \begin{cases} v'(z), & \text{if } z \in V', \\ a, & \text{if } z = x, \end{cases}$$

satisfies all equations in $S_C(x)$ and moreover $v(x) = a$, as required. $\square$

We are going to present a procedure that transforms each system of equations into a more regular one at the expense of adding some definable constraints. More formally, this `Simplify` procedure takes an instance I of SYSTERMSAT$(\mathbf{A})$ to remove equations during an iterative process. Finally it produces an instance I' of CSYSTERMSAT$(\mathbf{A})$ where a constraint function C is allowed and I' is supposed to satisfy:

(1) I' is equivalent to I,
(2) all equations in I' are of the form $q(x) = r(y)$, where $q, r \in Clo_1(\mathbf{A})$ and x, y are different variables. Both q and r are generic and $|q(C(x)) \cap r(C(y))| > 1$,
(3) $C(x) \in \mathcal{C}(\mathbf{A})$ and $|C(x)| > 1$ for all $x \in \mathrm{Var}(I')$.

The procedure also returns a boolean value, with `False` meaning that the instance I has no solutions at all. The `True` value does not say however that I has a solution but allows to replace I by its equivalent modification I'.

```
 1  Simplify( I )
 2      V := Var(I)
 3      C(x) := A for all x ∈ V
 4      I' := (I, C)
 5
 6      repeat
 7          if exists ( x ∈ V,  C(x) = Ø ) then return False
 8          for each e ∈ I' do
 9              if e is c ≈ d and c, d are different constants then
10                  return False
11              if e is c ≈ c and c is a constant then
12                  remove e from I'
13              if e is t(x) ≈ c then
14                  remove e from I'
15                  C(x) := C(x) ∩ t⁻¹(c)
16              if e is t(x) ≈ s(x) then
17                  remove e from I'
18                  C(x) := C(x) ∩ {a ∈ A : t(a) = s(a)}
19              if e is s(x) ≈ p(y) and x, y are different variables
20                  and p is a permutation then
21                  remove e from I'
22                  t := p⁻¹s
23                  replace all occurrences of y in I' with t(x)
24                  C(x) := C(x) ∩ t⁻¹(C(y))
25              if e is t(x) ≈ s(y) and x, y are different variables
26                  and t(C(x)) ∩ s(C(y)) = Ø then
27                  return False
28              if e is t(x) ≈ s(y) and x, y are different variables
29                  and t(C(x)) ∩ s(C(y)) = {c} for some c ∈ A then
30                  remove e from I'
31                  C(x) := C(x) ∩ t⁻¹(c)
32                  C(y) := C(y) ∩ s⁻¹(c)
33      until not changed( I' )
34
35      return True
```

Looking at what is done in the repeat loop we easily see that each time the instance I' is transformed to its equivalent form. Only a few words are needed to comment lines 19–24. The equation $s(x) \approx p(y)$ is equivalent to $p^{-1}s(x) \approx y$. Therefore introducing $t = p^{-1}s$ and replacing all occurrences of y by $t(x)$ together with updating the constraint for x by requiring $t(x) \in C(y)$ transforms I' to an equivalent form. Since I' is equivalent to I we get the answer for I by considering I'. Actually one can transform any solution of I' to a solution of I by remembering the replacements for variables from the set $V \setminus \text{Var}(I')$.

To see that (2) is satisfied observe that repeat loop removes all no-variable and one-variable equations. Suppose $q(x) \approx r(y)$ is a two-variables equation with an operation, say q, that is not generic. If q is a permutation then the equation $q(x) \approx r(y)$ is removed in line 21. Otherwise q is a constant. However then $|q(C(x)) \cap r(C(y))| \leqslant |q(C(x))| = 1$ thus either False is returned in line 27 or the equation is removed in line 30. Finally, all two-variables equation that are left in I' satisfy $|q(C(x)) \cap r(C(y))| > 1$ thanks to condition from line 29.

If I' does not satisfy (2) then each run of the repeat loop either returns False or removes at least one equation. Therefore there are at most $|I|$ runs of the loop. On the other hand each single run can be done in linear time so that Simplify works in quadratic time.

To prove that $|C(x)| > 1$, first pick a variable $x \in \text{Var}(I')$. Then observe that line 7 gives $|C(x)| \neq 0$. Suppose $|C(x)| = 1$ and take a two-variables equation from I' that contains x. We immediately get $|q(C(x)) \cap r(C(y))| \leqslant |q(C(x))| \leqslant |C(x)| = 1$ which is a contradiction to the property (2).

The rest of the last property (3) is covered by the following Claim.

Claim 2.17 *Whenever* Simplify *sets a value of the constraint function C for a variable x we have $C(x) \in \mathcal{C}(\mathbf{A})$. Moreover, if* Simplify *procedure returns* False *then $\emptyset \in \mathcal{C}(\mathbf{A})$.*

Proof We initialize the constraint function C in line 3 by putting the value A. Obviously $A = \{a \in A : id(a) = id(a)\} \in \mathcal{C}(\mathbf{A})$. During the main loop of Simplify we change values of the constraint function C in five cases. We are going to show that after every change $C(x) \in \mathcal{C}(\mathbf{A})$. First time we make a change in line 15 by putting $C(x) := C(x) \cap t^{-1}(c)$ so that we need to argue that $t^{-1}(c) \in \mathcal{C}(\mathbf{A})$. However $\{c\}$ is the image of A by the constant term operation $c \in Clo_1(\mathbf{A})$ and then $t^{-1}(c)$ is definable. Similar argument applies to line 18 with $C(x) := C(x) \cap \{a \in A : t(a) = s(a)\}$ and to line 24 with $C(x) := C(x) \cap t^{-1}(C(y))$. The situation in lines 31–32 is very similar to the one from line 15. Again we have to argue that $\{c\} \in \mathcal{C}(\mathbf{A})$. This time however, we do not know that c is a constant expressible by a term operation from $Clo_1(\mathbf{A})$. But we have $\{c\} = t(C(x)) \cap s(C(y))$, thus $\{c\} \in \mathcal{C}(\mathbf{A})$, as required.

The Simplify procedure returns False in three cases. First time it happens in line 7, after detecting x with $C(x) = \emptyset$. Since $\mathcal{C}(\mathbf{A})$ contains all sets of the form $C(x)$ we have $\emptyset \in \mathcal{C}(\mathbf{A})$, as required. Next, in line 9 there is an equation $c \approx d$ with $c \neq d$. However $\emptyset = \{a \in A : c(a) = d(a)\} \in \mathcal{C}(\mathbf{A})$. Finally False is returned in line 27, after detecting that $\emptyset = t(C(x)) \cap s(C(y))$ is the intersection of images of definable constraints. Thus $\emptyset \in \mathcal{C}(\mathbf{A})$.   □

2.2 Three-Element Algebras

The first example of a finite algebra $\mathbf{A}$ such that SYSPOLSAT$(\mathbf{A})$ is NP-complete goes back to Cook, when he had shown that SATISFIABILITY problem is NP-complete. This simply means that the boolean algebra $(2, \vee, \wedge, \neg)$ gives rise to NP-complete TERMSAT, POLSAT, SYSTERMSAT and SYSPOLSAT. Many other such examples can be derived from theorems of Chap. 1.

Our first lemma shows that SYSTERMSAT remains NP-complete even for some unary algebras (already with three elements).

Lemma 2.18 *There are finite unary algebras for which the problem* SYSTERMSAT (**A**) *is* NP-*complete.*

Proof A similar argument can be found in [FMS04]. Let $\mathbf{A} = (\{0, 1, 2\}, f, g, h)$ be an algebra with the following operations:

x	f	g	h
0	1	0	0
1	0	1	0
2	0	0	1

We have already mentioned in Chap. 1 that SYSTERMSAT (**A**) belongs to NP. To show that it is NP-complete we need the following version of the SATISFIABILITY problem:

POSITIVE-1-IN-3-SAT is a problem taking on its input a formula $F = C_1 \wedge \ldots \wedge C_n$, in which each clause C_i is of the form $(x \vee y \vee z)$, where x, y, z are (non-negated) variables, and answering the question if there is a boolean valuation such that in each clause *exactly one* variable takes value 1. For example for a formula $(x \vee y \vee z) \wedge (x \vee t \vee v) \wedge (v \vee t \vee z)$ the positive answer can be witnessed by $(0 \vee 1 \vee 0) \wedge (0 \vee 0 \vee 1) \wedge (1 \vee 0 \vee 0)$. It is easy to check by Schaefer [Sch78] result[2] (see also Garey and Johnson [GJ79]) that POSITIVE-1-IN-3-SAT is NP-complete.

Next we reduce POSITIVE-1-IN-3-SAT to systems of equations over the algebra **A**. A formula $F = C_1 \wedge \cdots \wedge C_n$ is transformed into $3n$ equations as follows:

$$C_i = x \vee y \vee z \quad \leadsto \quad \begin{cases} f(v_i) \approx x \\ g(v_i) \approx y \\ h(v_i) \approx z, \end{cases}$$

where $v_1, \ldots, v_n$ are new variables not occurring in F.

It is easy to check that a solution of the equations exists if and only if the formula F is satisfiable according to POSITIVE-1-IN-3-SAT rules. If v_i takes the value 0 (1 or 2) then only x (y or z respectively) takes boolean value 1 to make C_i true. $\square$

The following Lemma shows that the essence of the complexity of solving equations is hidden in generic operations:

Lemma 2.19 *For a unary algebra* $\mathbf{A} = (A, F)$ *with no generic operations in F the problem* SYSTERMSAT (**A**) *is in* P.

---

[2] The relation defined by POSITIVE-1-IN-3-SAT is $\{(1, 0, 0), (0, 1, 0), (0, 0, 1)\}$ and it is not closed under any of the operations listed in Fact 1.15.

Proof We apply `Simplify` procedure to an instance I. If we get `False` then there is no solution. Otherwise, because no operation in F, and thus in $Clo_1(\mathbf{A})$, is generic, the instance I' computed by `Simplify` has empty set of equations. Thus I' has a solution.   □

Since two-element unary algebras have no generic operations we immediately get from Lemma 2.19 that SYSTERMSAT over two-element unary algebras is in P.

For our main Theorem of this section we need to define the following:

Definition 2.20 We call a unary algebra $\mathbf{A} = (A, F)$ *proper* if there is no $a \in A$ with $f(a) = a$ for all $f \in Clo_1(\mathbf{A})$.

Consider SYSTERMSAT$(\mathbf{A})$ over a unary algebra $\mathbf{A} = (A, F)$ that is not proper, with $a \in A$ being a witness described in Definition 2.20. The valuation putting a for all variables obviously satisfies all term equations over $\mathbf{A}$. Thus the problem SYSTERMSAT$(\mathbf{A})$ can be solved in a constant time, with positive answer for all instances.

For a unary algebra $\mathbf{A} = (A, F)$ and $f \in Clo_1(\mathbf{A})$ by $Ker(f) = \{(x, y) \in A^2 : f(x) = f(y)\}$ we denote the *kernel* of f.

Definition 2.21 For a unary algebra $\mathbf{A} = (A, F)$ we define a *preorder* $\leqslant$ on $Clo_1(\mathbf{A})$ by putting $f \leqslant g$ if and only if $Ker(f) \subseteq Ker(g)$. We also put $\mathbf{P}(\mathbf{A}) = (Clo_1(\mathbf{A}), \leqslant)$.

Example 2.22 In the algebra $\mathbf{A} = (\{0, 1, 2\}, f, g, h)$, with the operations:

x	f	g	h
0	0	0	0
1	0	1	0
2	0	0	1

we have $g \leqslant f, h \leqslant f$ while g, h are incomparable.

By the width of an ordered set (or more generally, a preordered set) $\mathbf{P}$ we mean the largest number of pairwise incomparable elements of $\mathbf{P}$.

For our next theorem we need two lemmas:

Lemma 2.23 *For a proper unary algebra* $\mathbf{A}$ *with three elements the problem* SYSTERMSAT$(\mathbf{A})$ *is* NP*-complete if* width$(\mathbf{P}(\mathbf{A})) = 3$.

Proof To witness width 3 in the preorder $\mathbf{P}(\mathbf{A})$ the algebra $\mathbf{A}$ must have 3 term operations f_0, f_1, f_2 with pairwise incomparable kernels. Without loss of generality we may assume that f_0, f_1, f_2 act as follows ($a_i \neq b_i$):

x	f_0	f_1	f_2
0	b_0	a_1	a_2
1	a_0	b_1	a_2
2	a_0	a_1	b_2

We are going to show that there are $f, g, h \in Clo_1(\mathbf{A})$ and $\bot \neq \top$ in A such that either $P1$ or $P2$ holds:

	P1				P2		
x	f	g	h	x	f	g	h
0	$\top$	$\bot$	$\bot$	0	$\bot$	$\bot$	$\bot$
1	$\bot$	$\top$	$\bot$	1	$\top$	$\top$	$\bot$
2	$\bot$	$\bot$	$\top$	2	$\top$	$\bot$	$\top$

- Case 1: There exists $i \in \{0, 1, 2\}$ such that $i \neq b_i$.
  Without loss of generality we may assume that $i = 0$ and $b_0 = 1$. Then for $f_3 := f_1 f_0 \in Clo_1(\mathbf{A})$ we have:

x	f_0	f_1	f_2	f_3
0	1	a_1	a_2	b_1
1	a_0	b_1	a_2	a_1
2	a_0	a_1	b_2	a_1

  - Subcase 1.1: $\{a_1, b_1\} \neq \{0, 1\}$. Then $f_2 f_3$, $f_2 f_1$, f_2 satisfy:

x	$f_2 f_3$	$f_2 f_1$	f_2		x	$f_2 f_3$	$f_2 f_1$	f_2
0	b_2	a_2	a_2	or	0	a_2	b_2	a_2
1	a_2	b_2	a_2		1	b_2	a_2	a_2
2	a_2	a_2	b_2		2	b_2	b_2	b_2

  i.e., $P1$ or $P2$ (with 0 and 2 interchanged) holds.

  - Subcase 1.2: $\{a_1, b_1\} = \{0, 1\}$. Since we are not going to use f_0 any more, without loss of generality we may assume[3] that $a_1 = 0$, $b_1 = 1$. Observe that if $\{a_2, b_2\} = \{0, 1\}$ then f_3, f_1, f_2 satisfy either $P1$ or $P2$ (with 0 and 2 interchanged). Let $\{a_2, b_2\} \neq \{0, 1\}$ and put:

$$f_4 = \begin{cases} f_1 f_2, & \text{if } b_2 = 2 \text{ and } a_2 = 1, \\ f_3 f_2, & \text{if } b_2 = 2 \text{ and } a_2 = 0, \end{cases}$$

$$f_5 = \begin{cases} f_1 f_2, & \text{if } a_2 = 2 \text{ and } b_2 = 1, \\ f_3 f_2, & \text{if } a_2 = 2 \text{ and } b_2 = 0, \end{cases}$$

---

[3] We will use only f_1, f_2 and f_3 for which the situation $a_1 = 1$, $b_1 = 0$ is symmetric.

to get:

x	f_1	f_3	f_2	f_4	f_5
0	0	1	a_2	1	0
1	1	0	a_2	1	0
2	0	0	b_2	0	1

Thus f_3, f_1, f_4 satisfy $P2$ (with 0 and 2 interchanged) or f_3, f_1, f_5 satisfy $P1$.

- Case 2: For each $i \in \{0, 1, 2\}$ we have $i = b_i$.
  Without loss of generality we may assume that $a_0 = 1$, so that:

x	f_0	f_1	f_2
0	0	a_1	a_2
1	1	1	a_2
2	1	a_1	2

If $a_2 = 0$ then replacing f_2 by $f_0 f_2$ puts us into Case 1. If $a_2 = 1$ and $a_1 = 2$ then the term operations $f_1 f_0$, f_1, f_2 put us into $P2$ situation (with 0 and 1 interchanged). Finally, if $a_2 = 1$ and $a_1 = 0$ the term operations f_0, f_1, $f_1 f_2$ again put us into $P2$ situation (with 0 and 1 interchanged).

Now we know that **A** has 3 term operations f, g, h satisfying either $P1$ or $P2$.

Being in situation $P1$ we use the reduction of POSITIVE-1-IN-3-SAT presented in the proof of Lemma 2.18 to conclude that SYSTERMSAT(**A**) is NP-complete.

The reduction of POSITIVE-1-IN-3-SAT in situation $P2$ is only a bit harder. First we are going to show that if we are in $P2$ and not in $P1$ then a constant term operation $\top$ belongs to $Clo_1(\mathbf{A})$. If $\{\bot, \top\} = \{1, 2\}$ then $ff = \top \in Clo_1(\mathbf{A})$. Assume $\{\bot, \top\} \neq \{1, 2\}$.

- Case 1: $\top = 0$. In this situation ff, g, h satisfy $P1$.

x	f	g	h	ff
0	$\bot$	$\bot$	$\bot$	0
1	0	0	$\bot$	$\bot$
2	0	$\bot$	0	$\bot$

- Case 2: $\bot = 0$. Since **A** is proper, then there exists a term operation d such that $d(0) = d_0 \neq 0$. We analyze below the term operation fdf:

x	f	g	h	d	df	fdf
0	0	0	0	d_0	d_0	$\top$
1	$\top$	$\top$	0	d_1	d_3	d_4
2	$\top$	0	$\top$	d_2	d_3	d_4

If $d_4 = \top$ then $\top = fdf \in Clo_1(\mathbf{A})$. Otherwise $d_4 = 0$, thus fdf, g, h satisfy $P1$.

A formula $F = C_1 \wedge \cdots \wedge C_n$ is transformed into equations as follows. For each variable x occurring in F we need additional two variables v_x and x' and 3 equations:

$$ x \rightsquigarrow \begin{cases} f(v_x) \approx \top \\ g(v_x) \approx x \\ h(v_x) \approx x' \end{cases} $$

Next, for each clause C_i we need a variable v_i and then C_i is transformed into 3 equations as follows:

$$ C_i = x \vee y \vee z \quad \rightsquigarrow \quad \begin{cases} f(v_i) \approx x' \\ g(v_i) \approx y \\ h(v_i) \approx z \end{cases} $$

The equations for variable x force x' to simulate the negation of x. Indeed, because of the equation $f(v_x) \approx \top$ the variable v_x cannot be valuated to 0. If $v_x = 1$ then $x = \top$ and $x' = \bot$, while for $v_x = 2$ we have $x = \bot$ and $x' = \top$. The equations for the clause C_i work as in the proof of Lemma 2.18. Indeed, if $v_i = 0$ then x', y, z take value $\bot$ and thus x take value $\top$. If v_i takes value 1 or 2 again exactly one of the variables x, y, z takes value $\top$.

Lemma 2.24 *For a unary algebra* $\mathbf{A}$ *with three elements the problem* SYSTERM-SAT$(\mathbf{A})$ *is in* P *(in fact it is* $O(n^2)$*) if* width $(\mathbf{P}(\mathbf{A})) \leqslant 2$.

Proof Given an instance I of SYSTERMSAT$(\mathbf{A})$ we first apply Simplify procedure. As a result we get False, meaning that there is no solution of I, or an equivalent instance I' satisfying:

(1) All equations are of the form $f(x) \approx g(y)$, where $x \neq y$, $|f(A)| = 2$ and $f(A) = g(A)$,
(2) $|C(x)| > 1$ for each $x \in \mathrm{Var}(I')$.

Now we present the algorithm solving such simplified instance I'. We do a reduction into 2-SAT which is known to be polynomial (in fact $O(n^2)$, see e.g. Papadimitriou [Pap94]). We put $V' = \mathrm{Var}(I')$ and T to be the set of generic operations from $Clo_1(\mathbf{A})$. We define set of variables of 2-SAT by $V^* = \{X_t^x : x \in V' \text{ and } t \in T\}$.

Since $|t(A)| = 2$ for $t \in T$, so there is exactly one $c_t \in A$ with $|t^{-1}(t(c_t))| = 1$. This means that for every $a \in A$ we have:

$$ t(a) = t(c_t) \Leftrightarrow a = c_t. \tag{$\star$} $$

With the use of c_t our intended interpretation of 2-SAT variables can be described by:

$$ X_t^x \text{ is valuated by } 1 \Leftrightarrow x \text{ is valuated by } c_t. $$

We start our construction of the instance I^* of 2-SAT by transforming each equation of the form $f(x) \approx g(y)$ into two 2-SAT clauses[4]:

$$X_f^x \Leftrightarrow X_g^y, \quad \text{if } f(c_f) = g(c_g),$$

or $(\star\star)$

$$X_f^x \Leftrightarrow \neg X_g^y, \text{ otherwise.}$$

For term operations $f, g \in T$ and a variable x with $C(x) = A$ we add clauses which code the interaction between 2-SAT variables X_f^x and X_g^x:

$$X_f^x \Rightarrow X_g^x, \quad \text{if } c_f = c_g,$$

or $(\star\star\star)$

$$X_f^x \Rightarrow \neg X_g^x, \text{ if } c_f \neq c_g.$$

If $C(x) \neq A$ then from (2) we know that $|C(x)| = 2$. We finish our construction of I^* by adding the following clauses for each such variable x and $f, g \in T$:

$$
\begin{aligned}
&\neg X_f^x, &&\text{whenever } c_f \notin C(x), \\
&X_f^x \Leftrightarrow X_g^x, &&\text{if } c_f = c_g \text{ and } \{c_f, c_g\} \subseteq C(x), &&(\star\star\star\star)\\
&X_f^x \Leftrightarrow \neg X_g^x, &&\text{if } c_f \neq c_g \text{ and } \{c_f, c_g\} \subseteq C(x).
\end{aligned}
$$

Take a solution $v : V' \to A$ of I'. We follow our intended interpretation to define a valuation $s : V^* \to 2$ by:

$$
s(X_t^x) = \begin{cases} 1, & \text{if } v(x) = c_t, \\ 0, & \text{otherwise.} \end{cases}
$$

To see that s is a solution of I^* first take a clause $C \in I^*$ (in fact a pair of 2-SAT clauses) generated by $(\star\star)$ for an equation of the form $f(x) \approx g(y)$. We have $f(v(x)) = g(v(y))$ and two cases to consider:

- Case 1: $f(c_f) = g(c_g)$, so that $C = (X_f^x \Leftrightarrow X_g^y)$. Thanks to $(\star)$ the following equivalences hold:

$$s(X_f^x) = 1 \Leftrightarrow v(x) = c_f \Leftrightarrow f(v(x)) = f(c_f) \Leftrightarrow f(v(x)) = g(c_g) \Leftrightarrow$$
$$g(v(y)) = g(c_g) \Leftrightarrow v(y) = c_g \Leftrightarrow s(X_g^y) = 1,$$

as required.
- Case 2: $f(c_f) \neq g(c_g)$, so that $C = (X_f^x \Leftrightarrow \neg X_g^y)$. The proof is very similar to Case 1. Remember that $f(A) = g(A)$ and $|f(A)| = 2$. This time the following equivalences hold:

---

[4]Obviously $X \Leftrightarrow Y$ is a pair of 2-SAT clauses: $\neg X \vee Y$ and $X \vee \neg Y$.

$$s(X_f^x) = 1 \Leftrightarrow v(x) = c_f \Leftrightarrow f(v(x)) = f(c_f) \Leftrightarrow f(v(x)) \neq g(c_g) \Leftrightarrow$$

$$g(v(y)) \neq g(c_g) \Leftrightarrow v(y) \neq c_g \Leftrightarrow s(X_g^y) = 0,$$

as required.

The other clauses in I^* were generated either by (★★★) or by (★★★★). First choose a variable $x \in V'$ with $C(x) = A$ and take a clause C generated by (★★★) for $f, g \in T$ to get:

- Case 1: $c_f = c_g$, so that $C = (X_f^x \Rightarrow X_g^x)$. We have:

$$s(X_f^x) = 1 \Leftrightarrow v(x) = c_f \Rightarrow v(x) = c_g \Leftrightarrow s(X_g^x) = 1,$$

as required.
- Case 2: $c_f \neq c_g$, so that $C = (X_f^x \Rightarrow \neg X_g^x)$. This time we have:

$$s(X_f^x) = 1 \Leftrightarrow v(x) = c_f \Rightarrow v(x) \neq c_g \Leftrightarrow s(X_g^x) = 0,$$

as required.

Finally choose a variable $x \in V'$ with $|C(x)| = 2$. If $c_f \notin C(x)$ for some $f \in T$ then $v(x) \neq c_f$ and thus $s(X_f^x) = 0$ as determined by the first type of clauses from (★★★★). Suppose $\{c_f, c_g\} \subseteq C(x)$ to get:

- Case 1: $c_f = c_g$, so that $X_f^x \Leftrightarrow X_g^x \in I^*$. Similarly to the situation $C(x) = A$ the following equivalences hold:

$$s(X_f^x) = 1 \Leftrightarrow v(x) = c_f \Leftrightarrow v(x) = c_g \Leftrightarrow s(X_g^x) = 1,$$

as required.
- Case 2: $c_f \neq c_g$, so that $X_f^x \Leftrightarrow \neg X_g^x \in I^*$. This time thanks to $|C(x)| = 2$ we have $v(x) = c_f \Leftrightarrow v(x) \neq c_g$ thus:

$$s(X_f^x) = 1 \Leftrightarrow v(x) = c_f \Leftrightarrow v(x) \neq c_g \Leftrightarrow s(X_g^x) = 0,$$

as required.

Conversely, for a boolean valuation $s : V^* \to 2$ satisfying all clauses from I^* we define a valuation $v : V' \to A$ by:

$$v(x) = \begin{cases} c_t, & \text{for some } t \in T \text{ such that } s(X_t^x) = 1, \quad \text{if such } t \text{ exists,} \\ a, & \text{such that } a \neq c_t \text{ for all } t \in T, \qquad \text{otherwise.} \end{cases}$$

To see that v is well defined in the situation when $s(X_t^x) = 0$ for all $t \in T$ first assume $C(x) = A$. Since width $(\mathbf{P(A)}) \leqslant 2$, there is $a \in A$ with $a \neq c_t$ for all $t \in T$. Now assume $|C(x)| = 2$. Since s is a solution of I^*, then I^* cannot contain any clause of the form $X_f^x \Leftrightarrow \neg X_g^x$ generated by (★★★★). Thus $|\{c_t : t \in T$ and $c_t \in C(x)\}| \leqslant 1$, so there is $a \in C(x)$ such that $a \neq c_t$ for all $t \in T$, which we choose as a value for x.

Claim 2.25 *For each $X_t^x \in V^\star$ we have $s(X_t^x) = 1 \Leftrightarrow v(x) = c_t$.*

Proof Assume to the contrary that $s(X_t^x) = 1$ and $v(x) \neq c_t$. By definition of v there is $t' \in T$ such that $s(X_{t'}^x) = 1$ and $v(x) = c_{t'} \in C(x)$, so obviously $c_t \neq c_{t'}$. Since $s(X_t^x) = 1$ then $\neg X_t^x \notin I^\star$, so $c_t \in C(x)$ by (★★★★). Thus $\{c_t, c_{t'}\} \subseteq C(x)$ and by (★★★) or (★★★★) we get that the clause $X_{t'}^x \Rightarrow \neg X_t^x$ is in $I^\star$. We get a contradiction with $s(X_{t'}^x) = s(X_t^x) = 1$.

Now take $s(X_t^x) = 0$ to show $v(x) \neq c_t$. First observe that if $c_t \notin C(x)$ then obviously $v(x) \neq c_t$. Now assume $c_t \in C(x)$ to get two possibilities:

- Case 1: $s(X_{t'}^x) = 1$ for some $t' \in T$ and $v(x) = c_{t'} \in C(x)$ thus $\{c_{t'}, c_t\} \subseteq C(x)$. Since the clause $X_{t'}^x \Rightarrow X_t^x$ is not satisfiable then, by (★★★) or (★★★★), we get that $c_t \neq c_{t'}$ and thus $v(x) \neq c_t$, as required.
- Case 2: $s(X_{t'}^x) = 0$ for all $t' \in T$. By the definition of v we get $v(x) \neq c_t$.   □

To prove that v is a solution of I' take an equation of the form $f(x) \approx g(y)$ from I'.

- Case 1: $f(c_f) = g(c_g)$. Since s is a solution of $I^\star$ we get by (★★) that $s(X_f^x) = 1 \Leftrightarrow s(X_g^y) = 1$. Thanks to Claim 2.25 and (★) the following equivalences hold:

$$f(v(x)) = f(c_f) \Leftrightarrow v(x) = c_f \Leftrightarrow s(X_f^x) = 1 \Leftrightarrow s(X_g^y) = 1 \Leftrightarrow$$

$$v(y) = c_g \Leftrightarrow g(v(y)) = g(c_g) \Leftrightarrow g(v(y)) = f(c_f).$$

Thus $f(v(x)) = g(v(y))$ as $f(A) = g(A)$ has only two elements.

- Case 2: $f(c_f) \neq g(c_g)$, so that $s(X_f^x) = 1 \Leftrightarrow s(X_g^y) = 0$. The proof is very similar to Case 1. This time the following equivalences hold:

$$f(v(x)) = f(c_f) \Leftrightarrow v(x) = c_f \Leftrightarrow s(X_f^x) = 1 \Leftrightarrow s(X_g^y) = 0 \Leftrightarrow$$

$$v(y) \neq c_g \Leftrightarrow g(v(y)) \neq g(c_g) \Leftrightarrow g(v(y)) = f(c_f).$$

Thus $f(v(x)) = g(v(y))$ as required.   □

Now we are ready to state the main theorem of this section.

Theorem 2.26 [Bro06] *For a unary algebra $\mathbf{A}$ with at most three elements, SYSTERMSAT$(\mathbf{A})$ is in P (in fact it is $O(n^2)$) if $\mathbf{A}$ is not proper or width $(\mathbf{P}(\mathbf{A})) \leqslant 2$ holds, otherwise it is NP-complete.*

Proof Note that width $(\mathbf{P}(\mathbf{A})) = 1$ for any two-element algebra. On the other hand on the two-element set all four unary operations are not generic, thus Lemma 2.19 gives us that SYSTERMSAT$(\mathbf{A})$ is in P. For $|A| = 3$ we directly apply Lemmas 2.23 and 2.24.   □

2.3 Width and Complexity

We have seen in Theorem 2.26 that for three-element unary algebras $\mathbf{A}$ the complexity of SYSTERMSAT $(\mathbf{A})$ is fully characterized by the width of the corresponding preorder $\mathbf{P}(\mathbf{A})$.

Unfortunately this characterization does not extend to larger algebras. We are going to show that width and computational complexity are independent.

Observation 2.27 For the following algebra $\mathbf{A} = (A, F)$ with $F = \{f, g, h, r, s, 1\}$ defined by:

x	f	g	h	r	s	1
0	0	0	0	0	0	1
1	0	0	0	0	0	1
2	0	0	0	0	0	1
3	0	0	0	0	0	1
4	1	2	3	2	1	1
5	2	3	1	1	1	1
6	3	1	2	1	1	1

we have width $(\mathbf{P}(\mathbf{A})) = 1$ and SYSTERMSAT $(\mathbf{A})$ is NP-complete.

Proof First observe that $Clo_1(\mathbf{A}) = F \cup \{0, id\}$. Thus $\mathbf{P}(\mathbf{A})$ is a chain $Ker(id) \subseteq Ker(f) = Ker(g) = Ker(h) \subseteq Ker(r) \subseteq Ker(s) \subseteq Ker(1) = Ker(0)$, so that width $(\mathbf{P}(\mathbf{A})) = 1$.

Next consider the following system of term equations over $\mathbf{A}$:

$$\begin{cases} s(a) \approx 1 \\ f(a) \approx g(b) \\ f(a) \approx h(c) \end{cases}$$

and observe that there are only three possibilities for values of the quadruple $(a, r(a), r(b), r(c))$ in the solutions of the above system:

a	$r(a)$	$r(b)$	$r(c)$
4	2	1	1
5	1	2	1
6	1	1	2

This allows us to make a reduction from POSITIVE-1-IN-3-SAT similar to the one presented in Lemma 2.18. A formula $F = C_1 \wedge \cdots \wedge C_n$ is transformed into $6n$ equations as follows:

$$C_i = x \vee y \vee z \quad \rightsquigarrow \quad \begin{cases} s(a_i) \approx 1 \\ f(a_i) \approx g(b_i) \\ f(a_i) \approx h(c_i) \\ r(a_i) \approx x \\ r(b_i) \approx y \\ r(c_i) \approx z, \end{cases}$$

where a_i, b_i, c_i are new variables not occurring in F.

It is easy to check that a solution of these $6n$ equations exists if and only if the formula F is satisfiable according to POSITIVE-1-IN-3-SAT rules. If a_i takes the value 4 (5 or 6) then only x (y or z, respectively) is valuated by 2, while other two variables from $\{x, y, z\}$ are valuated by 1. Transforming 2 and 1 into true and false boolean values we make C_i true according to POSITIVE-1-IN-3-SAT rules.   □

To give examples of algebras **A** with arbitrarily large width (**P(A)**), but for which SYSTERMSAT(**A**) is solvable in a polynomial time, we need the following Lemma:

Lemma 2.28 *Let* $\mathbf{A} = (A, F)$ *be a unary algebra in which there is* $a_0 \in A$ *such that for all non-constant* $f \in F$:

(1) $f(a_0) = a_0$.
(2) $|f^{-1}(a)| \leqslant 1$ *for all* $a \neq a_0$.

Then SYSTERMSAT(**A**) *is in* P.

Proof Given an instance I of SYSTERMSAT(**A**) we apply the Simplify procedure to get an instance I' of CSYSTERMSAT(**A**). If Simplify procedure returns True then we valuate all variables in V by a_0. Observe that by (1) all generic operations $t \in Clo_1(\mathbf{A})$ satisfy $t(a_0) = a_0$. Since I' has only equations with generic operations, we get that all equations in I' are satisfied. One can check that for all definable constraints C such that $|C| > 1$ we have $a_0 \in C$, so that the above valuation is a solution of the instance I' of CSYSTERMSAT(**A**).   □

One consequence of Lemma 2.28 blocks a natural generalization of Theorem 2.26:

Observation 2.29 For each n there exists a proper algebra $\mathbf{A}_n$ such that width $(\mathbf{P}(\mathbf{A}_n)) = n$ and SYSTERMSAT$(\mathbf{A}_n)$ is in P.

Proof Put $\mathbf{A}_n = (\{0, \ldots, n\}, f_1, \ldots, f_n, c)$, where:

x	f_1	f_2	f_3	$\ldots$	f_n	c
0	0	0	0	$\ldots$	0	1
1	1	0	0	$\ldots$	0	1
2	0	1	0	$\ldots$	0	1
3	0	0	1	$\ldots$	0	1
$\vdots$	$\vdots$	$\vdots$	$\vdots$	$\ddots$	$\vdots$	$\vdots$
n	0	0	0	$\ldots$	1	1

The algebra A_n is proper and satisfies (1) and (2) of Lemma 2.28 with $a_0 = 0$. Therefore SYSTERMSAT (A_n) is in P. On the other hand one can easily check that width$(P(A_n)) = n$. □

Note however, that the problem CSYSTERMSAT (A_n) for $n \geqslant 3$ is NP-complete. It is because in CSYSTERMSAT we are allowed to use arbitrarily chosen constraints, e.g. of the form $C(x) = \{1, 2, 3\}$. Such constraint makes possible a reduction similar to the one from Lemma 2.18 (from POSITIVE-1-IN-3-SAT) with additional twist given by putting $C(v_i) = \{1, 2, 3\}$ for each variable v_i. Therefore we introduced the concept of *definable* constraints, which prevents us from such situations when translating instances of SYSTERMSAT into instances of CSYSTERMSAT in the `Simplify` procedure.

From Observation 2.27 we know that width 1 does not suffice to put SYSTERMSAT into P. However an assumption stronger than width 1, namely that the algebra A has exactly one non-constant operation is sufficient, as can be seen from the following Lemma:

Lemma 2.30 *For a unary algebra* $A = (A, F)$ *with exactly one non-constant operation in F the problem* SYSTERMSAT(A) *is in* P.

Proof Given an instance I of SYSTERMSAT(A) we apply the `Simplify` procedure to get an instance I' of CSYSTERMSAT(A). Denote by f the only non-constant operation in F. Since in I' we are left only with equations with generic operations, all of the operations in I' can be expressed by f^i, for some $i > 0$. Now pick a variable $x \in \text{Var}(I')$ and choose minimal m such that $f^m(x)$ can be found somewhere in I'. This means that $i \geqslant m$ whenever $f^i(x)$ occurs in I'. We replace every $f^i(x)$ with $f^{i-m}(x')$ for a new variable x' to get instance I''. We also copy constraint function from I' to I'' and then put $C(x') := f^m(C(x))$ in I''. It is easy to check that I' and I'' are equivalent and equal in size.

Now observe that at least one equation in I'' contains $id(x')$, since m was chosen to be minimal. Thus we can apply `Simplify` procedure once again, to get instance smaller than I' and repeat the above process. Whenever `Simplify` procedure returns `False` we know that there is no solution of the starting instance I. Otherwise we will end up with empty instance meaning that the solution exists. The number of steps is at most linear (in the size of I), so that the whole algorithm works in a polynomial time. □

References

[Bro06] Broniek P (2006) Solving equations over small unary algebras, Discrete Math Theoret Comput Sci Proc AF, 49–60
[FMS04] Feder Tomás, Madelaine Florent, Stewart Iain A (2004) Dichotomies for classes of homomorphism problems involving unary functions. Theoret Comput Sci 314(1–2):1–43
[GJ79] Garey MR, Johnson DS (1979) Computers and intractability. A guide to the theory of NP-completeness. W. H. Freeman and Co., San Francisco, California

[Pap94] Papadimitriou CH (1994) Computational complexity. Addison-Wesley Publishing Company, Reading, MA

[Sch78] Schaefer TJ (1978) The complexity of satisfiability problems, In: Conference record of the tenth annual ACM symposium on theory of computing (San Diego, California, 1978), ACM, New York, pp. 216–226

Chapter 3
Reducing CSP to SYSTERMSAT over Unary Algebras

Abstract We present the computational complexity equivalence of Constraint Satisfaction Problem and solving systems of equations over unary algebras. We give detailed description of the polynomial transformation used both ways. We also prove that solving system of equations over arbitrary finite algebra is polynomially equivalent to solving system of equations over specially constructed unary algebra. Finally we show that if P is not equal to NP then the class of unary algebras for which solving systems of equations is solvable in a polynomial time is not closed under homomorphic images.

We are going to prove that the problems CONSTRAINT SATISFACTION PROBLEM and SYSTERMSAT over unary algebras are equivalent. First we will treat the case of relational structures with exactly one relational symbol.

Definition 3.1 Let $\mathbf{D} = (D, R)$ be a proper relational structure with exactly one relational symbol R. Assume $R \subseteq D^r$ and all r-tuples of relation R are numbered from 1 to n, where $R[i]$ is the ith tuple. By $R[i]_k$ we denote the element at the kth position in the tuple $R[i]$. We define a unary algebra $\widetilde{\mathbf{D}}$:

- $A = \{\top, \bot, 1, \ldots, n\} \cup D$, where $\{\top, \bot, 1, \ldots, n\} \cap D = \emptyset$,
- $F = (c_1, \ldots, c_r, p, \top, \bot, id)$,
- $c_k(x) = \begin{cases} R[x]_k, & \text{if } 1 \leqslant x \leqslant n, \\ \bot, & \text{otherwise,} \end{cases}$
- $p(x) = \begin{cases} \top, & \text{if } 1 \leqslant x \leqslant n, \\ \bot, & \text{otherwise,} \end{cases}$
- $\widetilde{\mathbf{D}} = (A, F)$.

x	c_1	$\ldots$	c_k	$\ldots$	c_r	p
$\{\top, \bot\} \cup D$	$\bot$	$\ldots$	$\bot$	$\ldots$	$\bot$	$\bot$
1	$R[1]_1$	$\ldots$	$R[1]_k$	$\ldots$	$R[1]_r$	$\top$
$\vdots$	$\vdots$		$\vdots$		$\vdots$	$\vdots$
i	$R[i]_1$	$\ldots$	$R[i]_k$	$\ldots$	$R[i]_r$	$\top$
$\vdots$	$\vdots$		$\vdots$		$\vdots$	$\vdots$
n	$R[n]_1$	$\ldots$	$R[n]_k$	$\ldots$	$R[n]_r$	$\top$

© The Author(s) 2015
P. Broniek, *Computational Complexity of Solving Equation Systems*,
SpringerBriefs in Philosophy, DOI 10.1007/978-3-319-21750-5_3

If **D** is not proper then we define $\widetilde{\mathbf{D}}$ to be the unary algebra with one element and no operations.

It is easy to check, that all unary term operations over $\widetilde{\mathbf{D}}$ are members of F. The following table shows all possibilities of compositions fg with $f, g \in F$:

$f \backslash^g$	c_j	p	$\top$	$\bot$	id
c_i	$\bot$	$\bot$	$\bot$	$\bot$	c_i
p	$\bot$	$\bot$	$\bot$	$\bot$	p
$\top$	$\top$	$\top$	$\top$	$\top$	$\top$
$\bot$	$\bot$	$\bot$	$\bot$	$\bot$	$\bot$
id	c_j	p	$\top$	$\bot$	id

Lemma 3.2 *Let* $\mathbf{D} = (D, R)$ *be a relational structure with exactly one relational symbol. The problem* CSP(**D**) *is polynomially equivalent to* SYSTERMSAT($\widetilde{\mathbf{D}}$).

Proof We need to present polynomial transformations between CSP(**D**) and SYS-TERMSAT($\widetilde{\mathbf{D}}$) in both ways.

Suppose **D** is not proper and thus every instance of CSP(**D**) is satisfiable. We transform all instances of CSP(**D**) into the same instance of SYSTERMSAT($\widetilde{\mathbf{D}}$) that consists of exactly one equation $\{x = x\}$ which is satisfiable. Similarly every instance of SYSTERMSAT($\widetilde{\mathbf{D}}$) is also always satisfiable. Thus we transform all instances of the problem SYSTERMSAT($\widetilde{\mathbf{D}}$) into the same instance of CSP(**D**), that consists of one variable X and one constraint $C = (R, (X, \ldots, X))$ which is obviously satisfiable. Therefore, in the rest of the proof we assume **D** is proper and $R \subseteq D^r$.

For our reduction from CSP(**D**) to SYSTERMSAT($\widetilde{\mathbf{D}}$) we take an instance $I = (V, \{C_1, \ldots, C_m\})$ of CSP(**D**) and construct I^*, an instance of SYSTERMSAT($\widetilde{\mathbf{D}}$). For each constraint $C_i = (R, (X_1^i, \ldots, X_r^i))$ in I we create a variable x_i and add the equation $p(x_i) \approx \top$ to I^*.

If the variables X_k^i and X_l^j occurring in constraints C_i and C_j of I are the same, we add the equation $c_k(x_i) \approx c_l(x_j)$ to I^*.

This transformation can be done in a quadratic time, as for every constraint C_i we construct one one-sided equation of the form $p(x_i) \approx \top$ and at most $r^2 m$ two sided equations.

To prove that I and I^* are equivalent, up to existence of solutions, first take a solution $s : V \to D$ of I. Each constraint C_i is satisfied by an appropriate tuple, i.e. $(s(X_1^i), \ldots, s(X_r^i)) \in R$. Suppose this tuple is k_ith tuple $R[k_i]$ out of n tuples of R. We define a valuation $v : \{x_1, \ldots, x_m\} \to A$ by:

$$v(x_i) = k_i \in \{1, \ldots, n\}.$$

Obviously the equations of the form $p(x_i) \approx \top$ are satisfied. For equations of the form $c_k(x_i) \approx c_l(x_j)$, by the definition of our transformation, the constraints C_i and C_j share the same variable at positions k and l, respectively. Thus $R[v(x_i)]_k =$

$R[v(x_j)]_l$ and therefore $c_k(v(x_i)) = R[v(x_i)]_k = R[v(x_j)]_l = c_l(v(x_j))$ as required.

Now let $v : \{x_1, \ldots, x_m\} \to A$ be a solution of $I^\star$. The equation $p(x_i) \approx \top$ gives that $v(x_i) \in \{1, \ldots, n\}$. To see that $s : V \to D$ determined by $s(X_k^i) = R[v(x_i)]_k$ is well defined suppose that variables $X_k^i, X_l^j \in V$ are the same. But then there is an equation $c_k(x_i) \approx c_l(x_j)$ in $I^\star$ and therefore $s(X_k^i) = R[v(x_i)]_k = c_k(v(x_i)) = c_l(v(x_j)) = R[v(x_j)]_l = s(X_l^j)$. Now we know that the tuple $(s(X_1^i), \ldots, s(X_r^i))$ is equal to $(R[v(x_i)]_1, \ldots, R[v(x_i)]_r) = R[v(x_i)] \in R$ and thus s is a solution of I. This concludes our reduction of CSP($\mathbf{D}$) to SYSTERMSAT($\widetilde{\mathbf{D}}$).

Conversely, to reduce SYSTERMSAT($\widetilde{\mathbf{D}}$) to CSP($\mathbf{D}$) we divide our transformation of an instance J of SYSTERMSAT($\widetilde{\mathbf{D}}$) to the corresponding instance $J^\star$ of CSP($\mathbf{D}$) into two parts. First we reduce J to an equivalent instance J' of SYSTERMSAT($\widetilde{\mathbf{D}}$). Next we construct $J^\star$ from J'. The idea standing behind the second part is to reverse the transformation from I to $I^\star$. Unfortunately our instance J is not necessarily of the form $I^\star$ for some instance I of CSP($\mathbf{D}$). Introducing our intermediate step J' will help to resolve this problem.

For our transformation we need two special instances N and Y of CSP($\mathbf{D}$). The instance N consists of one variable X and one constraint $C = (R, (X, \ldots, X))$. Since $\mathbf{D}$ is proper, the instance N is not satisfiable. Similarly, the instance Y is defined by r different variables $X_1, \ldots, X_r$ and one constraint $C = (R, (X_1, \ldots, X_r))$. Since R is nonempty, the instance Y is satisfiable.

The following triangular tables show all possible forms of equations and divide them into eight classes (i)–(viii). Obviously the tables are triangular since we only consider one of two symmetric equations $t \approx s$ and $s \approx t$. We use classification of equations from Observation 2.6.

no-variable

	$\top$	$\perp$
$\top$	(i)	(v)
$\perp$	–	(i)

one-sided one-variable x

	$\top$	$\perp$
c_i	(v)	(vii)
p	(vi)	(vii)
id	(iii)	(iv)

two-sided one-variable x

	c_j	p	id
c_i	(i) for $i = j$ / (viii) for $i \neq j$	(vii)	(iv)
p	–	(i)	(iv)
id	–	–	(i)

two-sided two-variables x, y

x \ y	c_j	p	id
c_i	(viii)	(vii)	(ii)
p	–	(viii)	(ii)
id	–	–	(ii)

The following `FirstPart` procedure presents our transformation from J to J'. During the iterative process described in lines 3–15 we completely remove equations that belong to one of the four classes (i)–(iv). In every step of this iteration the instance is changed into an equivalent one, as we will explain in the descriptions of the classes (i)–(viii). After making replacements in lines 8, 11 and 13 we always transform equations to one of the forms listed in Observation 2.6.

```
 1 FirstPart( J )
 2   J" := J
 3   repeat
 4     for each e ∈ J" do
 5       case class(e) of
 6           (i): remove e from J"
 7          (ii): remove e from J"
 8               replace all occurrences of y in J" with
 9                 the other side of e (i.e. c_i(x) / p(x) / x )
10         (iii): remove e from J"
11               replace all occurrences of x in J" with ⊤
12          (iv): remove e from J"
13               replace all occurrences of x in J" with ⊥
14     end
15   until not changed( J" )
16
17   if exists ( e ∈ J" and class(e) = (v) ) then
18     return N
19   if not exists ( e ∈ J" and class(e) = (vi) ) then
20     return Y
21   if not MarkVariables( J" ) then
22     return N
23
24   J' is J" after removing all equations with variables
25     outside the set Marked
```

We present here the description of the classes:

  (i) the equation is always satisfiable.
 (ii) the equation implies that we can replace all the occurrences of y by $c_i(x)$, $p(x)$ or x, respectively,
(iii) the equation is equivalent to $x = \top$,
(iv) the equation is equivalent to $x = \bot$,
 (v) the equation is never satisfiable,
(vi) the equation is equivalent to $x \in \{1, \ldots, n\}$
(vii) the equation is equivalent to $x \in A \setminus \{1, \ldots, n\}$ or to $x, y \in A \setminus \{1, \ldots, n\}$ if the equation has two variables x, y,
(viii) the equations are of the form $p(x) \approx p(y)$, $c_i(x) \approx c_j(x)$ or $c_i(x) \approx c_j(y)$ and are left unchanged.

Observe that when the `repeat` loop 3–15 ends, the only equations that are left belong to classes (v)–(viii). In the next lines 17–20 we take care of two easy cases. If there is an equation from class (v) then the instance is not satisfiable and is transformed into N. We set $V'' = \mathrm{Var}(J'')$. The instance J'' now does not have equation from classes (i)–(v). If there is no equation from class (vi) in J'' then the valuation $v : V'' \to A$, that puts $v(x) = \bot$ for every variable $x \in V''$, is a solution of J'' since it satisfies all equations from classes (vii) and (viii).

At this point, after line 20, we are left with equations from classes (vi)–(viii) and we know that there is at least one equation from class (vi). Now we isolate variables occurring in equations from class (vi) with the help of MarkVariables procedure described below. Actually we isolate all variables that are forced to take values from the set $\{1, \ldots, n\}$.

```
 1   MarkVariables( J″ )
 2      Marked := ∅
 3      for each e ∈ J″ do
 4         if class(e) = (vi) then
 5            x := Var(e)
 6            Marked := Marked ∪ {x}
 7
 8      repeat
 9         for each e ∈ J″ do
10            if class(e) = (viii) then
11               {x, y} := Var(e)
12               if x ∈ Marked and y ∉ Marked then
13                  Marked := Marked ∪ {y}
14      until not changed(Marked)
15
16      if exists ( e ⊂ J″ and class(e) = (vii) and
17                  e has a marked variable ) then
18         return False
19      else
20         return True
```

We claimed that if v is a solution of J'' (the argument of the procedure MarkVariables in line 21 of FirstPart) and x is a marked variable then $v(x) \in \{1, \ldots, n\}$. To see this, first observe that if a variable x is marked in line 6 then $v(x) \in \{1, \ldots, n\}$ obviously holds. If a variable y is marked in line 13 then y has a companion x that is already marked and there is an equation of the form $c_i(x) \approx c_j(y)$ or $p(x) \approx p(y)$ in J''. Thus we get $v(y) \in \{1, \ldots, n\}$.

Moreover, if a variable that occurs in an equation from class (vii) is marked then the instance J'' has no solution. Therefore returning False in line 18 will later lead to the instance N in FirstPart procedure.

On the other hand if MarkVariables returns True then in each two-variables equation of J'' either both variables are marked or none of them is. This allows us to divide all equations of J'' into two sets:

- J''_M – the set of equations with all variables marked,
- J''_U – the set of equations with all variables unmarked.

We also know, that marked variables do not occur in an equation from class (vii), thus no such equation belongs to J''_M. Thus J''_M contains equations from classes (vi) or (viii) while J''_U from classes (vii) or (viii). Finally FirstPart constructs J' by taking equations only from J''_M and forgetting about J''_U. Therefore FirstPart returns the set J' consisting of equations with all variables marked and thus of class (vi) or (viii). We put $V' = \text{Var}(J')$ to get that V' is obviously equal to the set Marked.

To see that J' is equivalent to J it remains to show that J' and J'' are equivalent. Since $J' \subseteq J''$, all we need is to construct a solution j'' of J'' from a solution j' of J'. Remembering that $J' = J''_M = J''\backslash J''_U$ and that J''_M and J''_U have no common variables we may define $j'' : V'' \to A$ by:

$$j''(x) = \begin{cases} j'(x), & \text{if } x \in V', \\ \bot, & \text{otherwise.} \end{cases}$$

As j'' is exactly the same as j' on the set V' of marked variables then obviously all equations from J''_M are satisfied. Moreover every equation $e \in J''_U$ belongs either to class (vii) or (viii), and therefore e is satisfied by sending all its variables to $\bot$. We conclude that j'' is a solution of J''.

The second part is to transform J' to $J^\star$. First we enumerate variables in V' to get $V' = \{x_1, \ldots, x_m\}$. We reverse the idea described when passing from I to $I^\star$.

On the set $X = \{X_k^i : i = 1, \ldots, m \text{ and } k = 1, \ldots, r\}$ we define a relation $\sim$ by putting $X_k^i \sim X_l^j$ if the equation $c_k(x_i) \approx c_l(x_j)$ is in J'. Then the smallest equivalence relation containing $\sim$ is denoted by $\simeq$. The instance $J^\star$ consists of:

- a set of variables $V^\star = X/\simeq$,
- a set of constraints $\{C_1, \ldots, C_m\}$, where $C_i = (R, (Y_1^i, \ldots, Y_r^i))$ and $Y_k^i = [X_k^i]_\simeq$.

To prove the equivalence of J' and $J^\star$ first take a solution $s : V^\star \to D$ of $J^\star$. Each constraint C_i is satisfied by an appropriate tuple, i.e. $(s(Y_1^i), \ldots, s(Y_r^i)) \in R$. Suppose this tuple is k_ith tuple $R[k_i]$ of R. We define a valuation $v : V' \to A$ by putting $v(x_i) = k_i$. Obviously the equations from class (vi) of the form $p(x_i) \approx \top$ are satisfied. The equations of the form $p(x_i) \approx p(x_j)$ from class (viii) are also satisfied. For equations of the form $c_k(x_i) \approx c_l(x_j)$ (possibly with $i = j$), by the definition of our transformation, the constraints C_i and C_j share the same variable at positions k and l respectively, i.e., $Y_k^i = Y_l^j$. Thus $R[v(x_i)]_k = R[v(x_j)]_l$ and therefore $c_k(v(x_i)) = R[v(x_i)]_k = R[v(x_j)]_l = c_l(v(x_j))$ as required. Thus v is a solution of J'.

Now let $v : V' \to A$ be a solution of J'. Since each variable x_i is marked, we know that $v(x_i) \in \{1, \ldots, n\}$. To see that $s : V^\star \to D$ determined by $s(Y_k^i) = R[v(x_i)]_k$ is well defined suppose that variables $Y_k^i, Y_l^j \in V^\star$ are the same. It means $X_k^i \simeq X_l^j$, which is witnessed by a sequence $X_k^i \sim \ldots \sim X_l^j$. However, for each step $X_{k'}^{i'} \sim X_{l'}^{j'}$ in this sequence the equation $c_{k'}(x_{i'}) \approx c_{l'}(x_{j'})$ is in J' and therefore $s(X_{k'}^{i'}) = R[v(x_{i'})]_{k'} = c_{k'}(v(x_{i'})) = c_{l'}(v(x_{j'})) = R[v(x_{j'})]_{l'} = s(X_{l'}^{j'})$. Thus s

takes the same value on entire equivalence class of $\simeq$. The tuple $(s(Y_1^i), \ldots, s(Y_r^i)) = (R[v(x_i)]_1, \ldots, R[v(x_i)]_r) = R[v(x_i)]$ obviously lies in R so that s is a solution of J^*.

It remains to show that the construction of J^* from J is done in a polynomial time. The FirstPart procedure can be implemented to work in quadratic time. In every step of the iteration the size of the instance decreases, thus the number of iterations is at most linear. Every iteration step can also be implemented in linear time. The MarkVariables procedure can be implemented in quadratic time. During the second part of our transformation we only need to compute the smallest equivalence relation containing $\sim$, which can be easily done in quadratic time by using disjoint-sets data structure.   □

Definition 3.3 For $R \subseteq D^r, S \subseteq D^s$ we define $R \times S \subseteq D^{r+s}$ by $R \times S = \{(d_1, \ldots, d_{r+s}) : (d_1, \ldots, d_r) \in \mathrm{Rand}(d_{r+1}, \ldots, d_{r+s}) \in S\}$. For a finite relational structure $\mathbf{D} = (D, \mathcal{R})$, where $\mathcal{R} = (R_1, \ldots, R_n)$, we define $\mathbf{D}^{\times}$ by $(D, R_1 \times \ldots \times R_n)$.

Theorem 3.4 *For a relational structure $\mathbf{D} = (D, \mathcal{R})$ the problem* CSP($\mathbf{D}$) *is polynomially equivalent to* SYSTERMSAT($\widetilde{\mathbf{D}^{\times}}$)*, where $\mathbf{D}^{\times}$ is the unary algebra corresponding to the relational structure $\mathbf{D}^{\times}$.*

Proof Observe that for two nonempty relations R and S, the problems CSP(D, R, S) and CSP($D, R \times S$) are polynomially equivalent.

First we show the transformation from an instance $I = (V, \{C_i\}_i)$ of CSP(D, R, S) to an instance I^* of CSP($D, R \times S$). We translate each constraint with the relation R of the form $C_i = (R, (X_1^i, \ldots, X_r^i))$ to the constraint $C_i^* = (R \times S, (X_1^i, \ldots, X_r^i, Y_1^i, \ldots, Y_s^i))$, where $Y_k^i \notin V$ and $Y_k^i \neq Y_{k'}^{i'}$ for $(i, k) \neq (i', k')$. We translate constraints with relation S similarly. The equivalence of I and I^* is now easy to check remembering that R and S were supposed to be non-empty.

Conversely we translate an instance $J = (V, \{C_j\}_j)$ of CSP($D, R \times S$) to an instance J^* of CSP(D, R, S). For each constraint $C_j = (R \times S, (X_1^j, \ldots, X_{r+s}^j))$ we create two constraints in J^*, that is $C_{2j-1}^* = (R, (X_1^j, \ldots, X_r^j))$ and $C_{2j}^* = (S, (X_{r+1}^j, \ldots, X_{r+s}^j))$. As previously, the equivalence comes from the definition of $R \times S$.

Inducting on the number $|\mathcal{R}|$ of relations we now get that CSP($\mathbf{D}$) is polynomially equivalent to CSP($\mathbf{D}^{\times}$). Given a relational structure $\mathbf{D} = (D, \mathcal{R})$ we apply several $\times$ operations on relations from $\mathcal{R}$. The resulting $\mathbf{D}^{\times}$ has a single relation R, which size is possibly very large, actually $|R|$ may be exponential on $|\mathcal{R}|$. However neither $\mathcal{R}$ nor R comes into the inputs of corresponding problems CSP($\mathbf{D}$) or CSP($\mathbf{D}^{\times}$). With $\mathbf{D}^{\times}$ having only one relation we are in the setting of Lemma 3.2 to get the unary algebra $\widetilde{\mathbf{D}^{\times}}$ such that SYSTERMSAT($\widetilde{\mathbf{D}^{\times}}$) is polynomially equivalent to CSP($\mathbf{D}^{\times}$) and thus to CSP($\mathbf{D}$).

We summarize Theorems 3.4 and 1.26 in the following remark.

Remark 3.5 Each, even partial, characterization of computational complexity trans-
fers from CSP directly to SysTermSat over unary algebras. The same holds in the
other direction. In particular the Dichotomy Conjecture for CSP is equivalent to the
Dichotomy Conjecture for SysTermSat over unary algebras.

Another interesting consequence of Theorems 3.4 and 1.26 is the following:

Remark 3.6 Let $\mathbf{A}$ be an arbitrary finite algebra. Then SysTermSat($\mathbf{A}$) is polynomially equivalent to SysTermSat($\widehat{\mathbf{A}}$), where $\widehat{\mathbf{A}}$ is some unary algebra.

From Remark 3.6 we know that giving computational complexity characterization
of SysTermSat over unary algebras gives the characterization for all algebras. This
provides another motivation for studying unary algebras. On the other hand it also
shows how hard unary algebras might be.

It is worth mentioning, that Theorem 3.4 does not allow us to polynomially transfer
a solution of the Meta-Problem of computational complexity of SysTermSat over
unary algebras to the solution of CSP. In Meta-Problem the structure *is* a part of the
input. However, passing from $\mathbf{D}$ to $\mathbf{D}^{\times}$ is not polynomial in size of the structure $\mathbf{D}$.

Remark 3.5 leaves us with a little hope for getting the full computational complexity characterization for SysTermSat over unary algebras, as it is exactly that
hard like obtaining it for the whole Constraint Satisfaction Problem. On the
other hand, one of the obstacles for characterizing unary algebras $\mathbf{A}$ with SysTerm-
Sat($\mathbf{A}$) in P lies in the next Observation. We show there that SysTermSat cannot
be reduced to smaller structures, as it can happen that after such reduction it appears
to be harder. This situation stays in a great contrast to CSP, where actually such
reduction is often done, see [Bul06].

Observation 3.7 If $\mathsf{P} \neq \mathsf{NP}$ then the class of unary algebras $\{\mathbf{A} : \text{SysTermSat}(\mathbf{A})$
$\in \mathsf{P}\}$ is not closed under homomorphic images.

Proof Note that the seven-element algebra $\mathbf{H}$ given by:

x	f	g	h	p	c
0	0	0	0	0	1
1	0	0	0	0	1
2	0	0	0	0	1
3	0	0	0	0	1
4	1	0	0	1	1
5	0	1	0	2	1
6	0	0	1	3	1

satisfies (1) and (2) of Lemma 2.28, so that SysTermSat($\mathbf{H}$) $\in \mathsf{P}$.

On the other hand congruence Θ with the only nontrivial block being $\{1, 2, 3\}$
gives rise to the quotient algebra $\mathbf{H}' = \mathbf{H}/\Theta$:

x	f	g	h	p	c
0	0	0	0	0	1
1	0	0	0	0	1
4	1	0	0	1	1
5	0	1	0	1	1
6	0	0	1	1	1

Following the argument in the proof of Lemma 2.18 with additional twist given by equations of the form $p(v_i) = 1$ for each variable v_i, we conclude that SYSTERMSAT($\mathbf{H}'$) is NP-complete. □

Reference

[Bul06] Bulatov AA (2006) A dichotomy theorem for constraint satisfaction problems on a 3-element set. J ACM 53(1):66–120 (electronic)

Chapter 4
Partial Characterizations

Abstract We present several partial characterization theorems of computational complexity of solving equations over subclasses of unary algebras. We classify strongly-2-generated and strongly-3-valued unary algebras. We also classify 2-valued four-element unary algebras. Finally we present a diagram of subclasses of four-element algebras that have computational complexity dichotomies and describe subclasses that need further work.

At the end of Chap. 3 we observed that currently it seems to be hopeless to find complete classification of the computational complexity of SYSTERMSAT over unary algebras. Therefore we decided to attack the problem of classifying unary algebras by giving partial characterizations. Our approach considers algebras with generic operations taking only few values. In particular, thanks to the tools we worked out, we are able to cover a wide class of four-element algebras.

4.1 Generic Operations with Few Values

The following two theorems allow us to reduce SYSTERMSAT over arbitrary large unary algebras, which generic operations take only few values, to the CONSTRAINT SATISFACTION PROBLEM over domains with these few values. In particular for two or three values one can obtain a classification from Schaefer or Bulatov characterization, respectively.

Theorem 4.1 *Let* A *be a strongly-2-generated unary algebra. Then* SYSTERMSAT (A) *is polynomially equivalent to* CSP $(\widehat{A})$ *for some two-element relational structure* $\widehat{A}$.

Theorem 4.2 *Let* A *be a strongly-3-valued unary algebra. Then* SYSTERMSAT (A) *is polynomially equivalent to* CSP $(\widehat{A})$ *for some three-element relational structure* $\widehat{A}$.

We decided to present one unified proof for both Theorems 4.1 and 4.2.

Proof Let $A = (A, F)$. We set $n = 2$ if A is a strongly-2-generated unary algebra or else $n = 3$ if A is a strongly-3-valued unary algebra.

Suppose $n = 2$ and $Clo_1(A)$ is generated by a strongly-2-valued set. Choose maximal, with respect to inclusion, strongly-2-valued G such that $Clo_1(A) =$

© The Author(s) 2015

P. Broniek, *Computational Complexity of Solving Equation Systems*,
SpringerBriefs in Philosophy, DOI 10.1007/978-3-319-21750-5_4

$Clo_1(A, G)$. Let T be the set of all generic operations from G. If $n = 3$ then, by our assumption, entire $Clo_1(\mathbf{A})$ is a strongly-3-valued set. In this case by T we mean the set of all generic operations from $Clo_1(\mathbf{A})$. Finally, there is $D = \{d_1, \ldots, d_n\}$ such that $t(A) \subseteq D$ for every operation $t \in T$. We also put $m = |T|$ so that T is listed by $(t_1, \ldots, t_m)$.

We define a function $\varepsilon : A \to D^m$ by $\varepsilon(a)_i = t_i(a)$. With the help of ε, for every $C \in \mathcal{C}(\mathbf{A})$ we define one m-ary relation R_C on D by putting:

$$R_C = \{\varepsilon(a) \in D^m : a \in C\} = \{(t_1(a), \ldots, t_m(a)) : a \in C\}.$$

Next we define a set $\mathcal{R}_0$ that consists of all relations of the form $R_C \subseteq D^m$, where $C \in \mathcal{C}(\mathbf{A})$ and $|C| > 1$. Finally we put:

$$\mathcal{R} = \begin{cases} \mathcal{R}_0 \cup \{R_=\}, & \text{if the relational structure } (D, \mathcal{R}_0) \text{ is proper} \\ & \text{or } \emptyset \notin \mathcal{C}(\mathbf{A}), \\ \{R_\perp\}, & \text{otherwise,} \end{cases}$$

where $R_=$ is the binary equality relation on D and $R_\perp = \{(d_1, \ldots, d_n)\}$. The relational structure we need is defined by $\widehat{\mathbf{A}} = (D, \mathcal{R})$.

To prove polynomial equivalence of the problems SYSTERMSAT$(\mathbf{A})$ and CSP$(\widehat{\mathbf{A}})$ we start with an instance I of SYSTERMSAT$(\mathbf{A})$. At the beginning we apply Simplify procedure.

Suppose Simplify returns False, meaning that there is no solution of the instance I. All we need to do here is to produce an unsatisfiable instance of CSP$(\widehat{\mathbf{A}})$. Claim 2.17 gives us that $\emptyset \in \mathcal{C}(\mathbf{A})$. If $\mathcal{R} = \{R_\perp\}$ then we return an instance of CSP$(\widehat{\mathbf{A}})$ that consists of one variable X and one constraint $C = (R_\perp, (X, \ldots, X))$, which is obviously not satisfiable. Now suppose $\mathcal{R} \neq \{R_\perp\}$ which together with $\emptyset \in \mathcal{C}(\mathbf{A})$ means that $(D, \mathcal{R}_0)$ is proper. We return an instance of CSP$(\widehat{\mathbf{A}})$ that consists of one variable X and $|\mathcal{R}_0|$ constraints of the form $(R_C, (X, \ldots, X))$, for each $R_C \in \mathcal{R}_0$. Since $(D, \mathcal{R}_0)$ is proper, the instance is not satisfiable.

If Simplify returns True then we get an instance I' of the problem CSYS-TERMSAT$(\mathbf{A})$. Suppose I' has no equations so that we need to produce a satisfiable instance of CSP$(\widehat{\mathbf{A}})$. If $\mathcal{R} = \{R_\perp\}$ then we return an instance of CSP$(\widehat{\mathbf{A}})$ that consists of n different variables $X_1, \ldots, X_n$ and one constraint $C = (R_\perp, (X_1, \ldots, X_n))$, which is obviously satisfiable by $(d_1, \ldots, d_n)$. Now consider $\mathcal{R} \neq \{R_\perp\}$ so that $R_= \in \mathcal{R}$ and the instance of CSP$(\widehat{\mathbf{A}})$ that consists of one variable X and one constraint of the form $C = (R_=, (X, X))$ is obviously satisfiable.

After cleaning our setting from easy cases we are left with a nonempty I' that has all equations of the form $q(x) \approx r(y)$, where $q, r \in Clo_1(\mathbf{A})$ and x, y are different variables. Moreover we know that $|q(C(x)) \cap r(C(y))| > 1$ and both q and r are generic.

If $n = 3$ then we are already left with operations q, r lying in T. Since for $n = 2$ our assumptions are weaker, we need a little more effort to reduce to the similar setting. As G generates q, we know that $q = g_1 \ldots g_l$ for some $g_1, \ldots, g_l \in G$. Since q is generic we know that none of g_i is a constant and there is g_i that is not a permutation. Choose minimal i such that g_i is not a permutation and put $p_q = g_1 \ldots g_{i-1}$ (if $i = 1$ then $p_q = id$) and $t_q = g_i \ldots g_l$ to get $q = p_q t_q$. Observe that p_q is a permutation. Since g_i is generic and belongs to G we have $g_i(A) \subseteq D$ so that $t_q(A) \subseteq D$. It follows that $G' = G \cup \{t_q\}$ is also strongly-2-valued. Since t_q is generated by G we also have $Clo_1(A, G) = Clo_1(A, G')$. As G was chosen to be maximal, $t_q \in G$. Actually $t_q \in T$ since t_q is generic. Analogously we decompose $r = p_r t_r$ into a permutation p_r and $t_r \in T$. Therefore $q(x) \approx r(y)$ is equivalent to $p_q t_q(x) \approx p_r t_r(y)$ or in other words to $t_q(x) \approx p' t_r(y)$ for the permutation $p' = p_q^{-1} p_r \in Clo_1(A)$. Moreover $\left| t_q(C(x)) \cap p' t_r(C(y)) \right| = \left| p_q^{-1}(p_q t_q(C(x)) \cap p_r t_r(C(y))) \right| = \left| p_q^{-1}(q(C(x)) \cap r(C(y))) \right| = |q(C(x)) \cap r (C(y))| > 1$. However both t_q and $p' t_r$ are 2-valued, therefore $t_q(A) = p' t_r(A)$. Consequently $p' t_r(A) = t_q(A) \subseteq D$, and we may argue that $p' t_r \in T$ by using maximality of G. We replace $q(x) \approx r(y)$ with $t_q(x) \approx p' t_r(y)$ to conclude that all equations in I' use only operations from T.

Let $V' := \mathrm{Var}(I') = \{x_1, \ldots, x_{v'}\}$. We construct an instance I^* of $\mathrm{CSP}(\widehat{A})$ consisting of:

- a set of variables $V = \left\{ X_j^i : x_i \in V' \text{ and } t_j \in T \right\}$,
- a set of constraints of two types:

  - $(R_=, (X_j^i, X_l^k))$, for each equation $t_j(x_i) \approx t_l(x_k)$ from I',
  - $(R_{C(x_i)}, (X_1^i, \ldots, X_m^i))$, for each variable $x_i \in V'$.

Since the constraint function C, and therefore the sets $C(x_i)$, are computed by the Simplify procedure, we know that $C(x_i) \in \mathcal{C}(A)$ and $|C(x_i)| > 1$ so that the relation $R_{C(x_i)} \in \mathcal{R}_0$.

To prove that I^* is equivalent to I' we start with a solution $v : V' \to A$ of I' and define a valuation $s : V \to D$ by:

$$s(X_j^i) = t_j(v(x_i)).$$

Consider a constraint of the form $(R_=, (X_j^i, X_l^k))$. Since v is a solution of I' and the equation $t_j(x_i) \approx t_l(x_k)$ belongs to I', we get $s(X_j^i) = t_j(v(x_i)) = t_l(v(x_k)) = s(X_l^k)$, as required. Now consider a constraint of the form $(R_{C(x_i)}, (X_1^i, \ldots, X_m^i))$. Observe, that for all $t_j \in T$ we have $\varepsilon(v(x_i))_j = t_j(v(x_i)) = s(X_j^i)$. We also have $v(x_i) \in C(x_i)$, thus $(s(X_1^i), \ldots, s(X_m^i)) = (\varepsilon(v(x_i))_1, \ldots, \varepsilon(v(x_i))_m) = \varepsilon(v(x_i)) \in R_{C(x_i)}$, as required.

Conversely, assume $s : V \to D$ is a solution of I^* and pick a variable $x_i \in V'$. Since the constraint $(R_{C(x_i)}, (X_1^i \ldots, X_m^i))$ is satisfied by s we have $(s(X_1^i), \ldots, s(X_m^i)) \in R_{C(x_i)}$. By the definition of $R_{C(x_i)}$ there is $a_i \in C(x_i)$ such that $\varepsilon(a_i) = (s(X_1^i), \ldots, s(X_m^i))$. Define a valuation $v : V' \to A$ by putting $v(x_i) = a_i$. Thus $v(x_i) \in C(x_i)$ as required by the constraint function C. Now consider an equation $t_j(x_i) \approx t_l(x_k)$ from I'. Since $(R_=, (X_j^i, X_l^k))$ is satisfied then $\varepsilon(v(x_i))_j = s(X_j^i) = s(X_l^k) = \varepsilon(v(x_k))_l$. It follows that $t_j(v(x_i)) = \varepsilon(v(x_i))_j =$

$\varepsilon(v(x_k))_l = t_l(v(x_k))$ so that the equation $t_j(x_i) \approx t_l(x_k)$ is satisfied. This finishes the proof that each instance I of SYSTERMSAT $(\mathbf{A})$ is equivalent to some instance $I^\star$ of CSP $(\widehat{\mathbf{A}})$.

Now take an instance J of CSP $(\widehat{\mathbf{A}})$ to produce an equivalent instance $J^\star$ of SYSTERMSAT $(\mathbf{A})$. Suppose first that $\mathcal{R} = \{R_\perp\}$ so that $\emptyset \in \mathcal{C}(\mathbf{A})$. Since $R_\perp$ contains only one tuple, every constraint of J has at most one solution. To determine whether J is satisfiable it is enough to check that these possible solutions do agree, which can be easily done in linear time. Suppose J is not satisfiable. Since $\emptyset \in \mathcal{C}(\mathbf{A})$, by Lemma 2.16 there is a set of equations $S_\emptyset(x)$ of SYSTERMSAT $(\mathbf{A})$ without a solution. We return $S_\emptyset(x)$ as $J^\star$. If J is satisfiable then we return an instance $J^\star$ that consists of exactly one equation $\{x = x\}$ and is obviously satisfiable.

Now assume $\mathcal{R} \neq \{R_\perp\}$. First with the help of $R_=$ we transform J to its equivalent form J' in which all variables in the constraints determined by a relation from $\mathcal{R}_0$ are different, i.e. $(R_C, (X_1, \ldots, X_m))$ has exactly m different variables while the sets of variables of the constraints $(R_{C_1}, (X_1, \ldots, X_m))$ and $(R_{C_2}, (Y_1, \ldots, Y_m))$ are disjoint. Indeed, e.g.:

$$(R_C, (X_1, X_1, X_3, \ldots, X_m))$$

can be replaced by:

$$\{(R_C, (X_1, X_2, X_3, \ldots, X_m)), (R_=, (X_1, X_2))\},$$

while:

$$\{(R_{C_1}, (X_1, X_2, \ldots, X_m)), (R_{C_2}, (X_1, Y_2, \ldots, Y_m))\}$$

by:

$$\{(R_{C_1}, (X_1, X_2, \ldots, X_m)), (R_{C_2}, (Y_1, Y_2, \ldots, Y_m)), (R_=, (X_1, Y_1))\}.$$

Obviously only a linear number of new constraints is added.

We denote by V' the set of variables of J'. Next we denote by $V_0 \subseteq V'$ the set of variables that occur in a constraint with relation from the set $\mathcal{R}_0$ and we put $V_1 = V' \setminus V_0$. We also define relation $\simeq$ over V' to be the smallest equivalence relation containing all pairs of the form $X \simeq Y$ whenever $(R_=, (X, Y)) \in J'$. Next we extend J' with constraints of the form $(R_=, (X, Y))$ for $X \simeq Y$. By changing notation the extended instance J' is obviously equivalent to the original one and contains at most quadratic (in terms of $|V'|$) number of new constraints. Finally we construct an instance J'' by removing all variables that belong to V_1 and all constraints with them. The set V'' of variables of J'' is obviously equal to V_0.

To prove that J'' and J' are equivalent we only need to show that every solution j'' of J'' can be extended to a solution j' of J'. Take a variable $X \notin V_0$. If an equivalence class $[X]_\simeq$ of X does not contain any variable from V_0 then we valuate, by j', all members of $[X]_\simeq$ to some arbitrarily chosen $d \in D$ to satisfy all constraints over variables from $[X]_\simeq$. On the other hand if $[X]_\simeq \cap V_0 \neq \emptyset$ then j'' sends entire $[X]_\simeq \cap V_0$ to a common value, say $d \in D$. Putting $j'(X') = d$ for all $X' \in [X]_\simeq \setminus V_0$ satisfies all constraints with relation $R_=$ and variables from the set $[X]_\simeq$.

Now, let C'' be the set of constraints from J'' that do not contain $R_=$. By our reduction from J to J' no variable occurs twice in C''. Moreover, thanks to reduction from J' to J'', every variable of J'' actually occurs in a constraint from C''. Thus each constraint $C_i \in C''$ is of the form $C_i = (R_{C^i}, (X_1^i, \ldots, X_m^i))$ with $C^i \in C(\mathbf{A})$. Hence $X_j^i \neq X_{j'}^{i'}$ for $(i, j) \neq (i', j')$ and in fact $V'' = \left\{ X_j^i : C_i \in C'' \text{ and } t_j \in T \right\}$.

Now we are ready to construct an instance J^* of SYSTERMSAT$(\mathbf{A})$. For every constraint $C_i = (R_{C^i}, (X_1^i, \ldots, X_m^i)) \in C''$ we create a variable x_i and add to J^* a set $S_{C^i}(x_i)$ of equations, defined by Lemma 2.16. We arrange these equations so that the $S_{C^i}(x_i)$'s have no common variables. Let $V^* = \bigcup_i \text{Var}(S_{C^i}(x_i))$. We also transform each constraint from J'' of the form $(R_=, (X_j^i, X_l^k))$ to an equation $t_j(x_i) \approx t_l(x_k)$ and add it to J^*.

To prove that J^* and J'' are equivalent, first take a solution $s : V'' \to D$ of J''. Every constraint $C_i = (R_{C^i}, (X_1^i, \ldots, X_m^i)) \in C''$ is satisfied by an appropriate tuple $\varepsilon(a_i) \in R_{C^i}$ for $a_i \in C^i$. Now, by Lemma 2.16, there is a solution v_i of $S_{C^i}(x_i)$ such that $v_i(x_i) = a_i$. By our arrangement no two sets of the form $S_{C^i}(x_i)$ have variables in common so that we can glue the v_i's to get a valuation $v : V^* \to A$.

We need to show that v is a solution of J^*. From the construction of v we know, that v satisfies all equations from all the $S_{C^i}(x_i)$'s. The other equations have the form $t_j(x_i) \approx t_l(x_k)$ coming from constraints $(R_=, (X_j^i, X_l^k))$ in J''. Thanks to our numeration of variables in V'' we know that X_j^i is placed in jth position in a constraint C_i, thus $s(X_j^i) = \varepsilon(a_i)_j = t_j(a_i) = t_j(v(x_i))$. Analogously $s(X_l^k) = \varepsilon(a_k)_l = t_l(a_k) = t_l(v(x_k))$. Since s satisfies the constraint $(R_=, (X_j^i, X_l^k))$, we get $s(X_j^i) = s(X_l^k)$. Finally $t_j(v(x_i)) = s(X_j^i) = s(X_l^k) = t_l(v(x_k))$, as required.

Conversely, take a solution $v : V^* \to A$ of J^*. Since v is a solution of the set $S_{C^i}(x_i)$, by Lemma 2.16, $v(x_i) \in C^i$ which gives $\varepsilon(v(x_i)) \in R_{C^i}$. We define a valuation $s : V'' \to D$ by putting $s(X_j^i) = \varepsilon(v(x_i))_j = t_j(v(x_i))$. Then $(s(X_1^i), \ldots, s(X_m^i)) = (\varepsilon(v(x_i))_1, \ldots, \varepsilon(v(x_i))_m) = \varepsilon(v(x_i)) \in R_{C^i}$. Thus the constraint C_i is satisfied by s, so that all constraints in C'' are. The remaining constraints from J'' are of the form $(R_=, (X_j^i, X_l^k))$. By construction we know that the equation $t_j(x_i) \approx t_l(x_k)$ belongs to J^*, thus $t_j(v(x_i)) = t_l(v(x_k))$. Finally $s(X_j^i) = t_j(v(x_i)) = t_l(v(x_k)) = s(X_l^k)$, as required. $\square$

The computational complexity of SYSTERMSAT$(\mathbf{A})$ for a strongly-2-generated unary algebra $\mathbf{A}$ and for a strongly-3-valued unary algebra $\mathbf{A}$ can be now characterized by the results of Schaefer [Sch78] and Bulatov [Bul06], respectively. This confirms the Dichotomy Conjecture 1.21 in these cases.

Corollary 4.3 *If $\mathbf{A}$ is a strongly-2-generated or a strongly-3-valued unary algebra then Dichotomy Conjecture holds for* SYSTERMSAT$(\mathbf{A})$.

4.2 Four-Element 2-Valued Algebras

The tools we have worked out up to now allow us to carefully analyze the situation
of four-element algebras in which all operations take 1, 2 or 4 values. In particular
the NP-complete situations caused (in Lemma 2.23) by width 3 in three-element
algebras can be generalized into arbitrarily large setting:

Lemma 4.4 *Let* $\mathbf{A} = (A, F)$ *be a unary algebra,* $A_1, A_2, A_3 \subseteq A$ *and* $f_1, f_2, f_3 \in$
$Clo_1(\mathbf{A})$ *with the following properties:*

- A_1, A_2 *and* A_3 *are nonempty and pairwise disjoint,*
- $f_i(A_i) = \{\top_i\}$ *for* $1 \leqslant i \leqslant 3$,
- $f_i(A_j) = \{\bot_i\}$ *for* $1 \leqslant i, j \leqslant 3$ *and* $i \neq j$,
- $\top_i \neq \bot_i$ *for* $1 \leqslant i \leqslant 3$,
- $A_1 \cup A_2 \cup A_3 \in \mathcal{C}(\mathbf{A}_A)$.

then SYSTERMSAT($\mathbf{A}_A$) *is* NP-*complete.*

Proof The properties of $\mathbf{A}$ can be described by the following table:

$x \in$	f_1	f_2	f_3
A_1	$\top_1$	$\bot_2$	$\bot_3$
A_2	$\bot_1$	$\top_2$	$\bot_3$
A_3	$\bot_1$	$\bot_2$	$\top_3$
$A \backslash (A_1 \cup A_2 \cup A_3)$		Arbitrary	

We are going to reduce the problem POSITIVE- 1- IN- 3- SAT to SYSTERMSAT($\mathbf{A}_A$).
Take a formula $F = C_1 \wedge \cdots \wedge C_n$, where $C_i = X_1^i \vee X_2^i \vee X_3^i$. We start the construction of an instance $F^\star$ of the problem SYSTERMSAT($\mathbf{A}_A$) by creating variables
$x_1, \ldots, x_n$, one for each clause C_i. If two variables X_α^i and X_β^j happen to be the same
we need to express this fact by some equations in $F^\star$. In particular if $X_\alpha^i = X_\alpha^j$ we put
into $F^\star$ the equation $f_\alpha(x_i) \approx f_\alpha(x_j)$. To treat the case $X_\alpha^i = X_\beta^j$ with $\alpha \neq \beta$ first we
let $Z = \left\{ (i, j, \alpha, \beta) : X_\alpha^i = X_\beta^j \text{ and } \alpha \neq \beta \right\}$. Now for each $k = (i, j, \alpha, \beta) \in Z$ we
create two completely new variables y_k and z_k and we include into $F^\star$ the following
set of equations:

$$\begin{cases} f_\alpha(x_i) \approx f_\alpha(y_k) \\ f_\beta(x_j) \approx f_\beta(z_k) \\ f_\gamma(y_k) \approx f_\gamma(z_k) \\ f_\alpha(z_k) \approx \bot_\alpha \\ f_\beta(y_k) \approx \bot_\beta, \end{cases} \qquad\qquad (\star)$$

where $\gamma \in \{1, 2, 3\} \setminus \{\alpha, \beta\}$.

Let $A' = A_1 \cup A_2 \cup A_3$ and V be the set of all created variables (i.e. all of the
x_i's, y_k's and z_k's). Finally, for each variable $x \in V$, we add the set of equations
$S_{A'}(x)$ to $F^\star$ supplied by Lemma 2.16 thanks to $A' \in \mathcal{C}(\mathbf{A}_A)$.

This transformation can be easily done in a polynomial time. We create at most quadratic number of variables and equations.

Let X be the set of all variables occurring in F and let $V^* = \mathrm{Var}(F^*)$. Now take a valuation $s : X \to \{0, 1\}$ satisfying formula F to produce a valuation $v : V^* \to A$. Since F^* contains the sets of the form $S_{A'}(x)$ for each $x \in V$, it is enough to valuate x by some element from A' to satisfy all of the $S_{A'}(x)$'s. To determine $v(x_i)$ note that since s is a POSITIVE- 1- IN- 3- SAT solution, we know that for each i exactly one of $s(X_1^i), s(X_2^i), s(X_3^i)$, say $s(X_\alpha^i)$, takes the true value. We set $v(x_i)$ to an arbitrarily chosen element of A_α. Whenever $s(X_\alpha^i) = s(X_\beta^j)$ this valuation v satisfies:

$$f_\alpha(v(x_i)) = \mathsf{T}_\alpha \Leftrightarrow s(X_\alpha^i) = 1 \Leftrightarrow s(X_\beta^j) = 1 \Leftrightarrow f_\beta(v(x_j)) = \mathsf{T}_\beta. \qquad (\star\star)$$

To determine $v(y_k)$ and $v(z_k)$ when $k = (i, j, \alpha, \beta) \in Z$, observe that $X_\alpha^i = X_\beta^j$ gives $s(X_\alpha^i) = s(X_\beta^j)$. Thus, by $(\star\star)$, we know that the pair $(f_\alpha(v(x_i)), f_\beta(v(x_j)))$ has to be either $(\mathsf{T}_\alpha, \mathsf{T}_\beta)$ or $(\perp_\alpha, \perp_\beta)$. However, one can check that for the sets of equations arising from these two cases:

$$\begin{cases} \mathsf{T}_\alpha \approx f_\alpha(y_k) \\ \mathsf{T}_\beta \approx f_\beta(z_k) \\ f_\gamma(y_k) \approx f_\gamma(z_k) \\ f_\alpha(z_k) \approx \perp_\alpha \\ f_\beta(y_k) \approx \perp_\beta, \end{cases} \qquad \begin{cases} \perp_\alpha \approx f_\alpha(y_k) \\ \perp_\beta \approx f_\beta(z_k) \\ f_\gamma(y_k) \approx f_\gamma(z_k) \\ f_\alpha(z_k) \approx \perp_\alpha \\ f_\beta(y_k) \approx \perp_\beta, \end{cases}$$

any pair from $A_\alpha \times A_\beta$ or $A_\gamma \times A_\gamma$, respectively, can serve as a solution for (y_k, z_k) satisfying $v(y_k), v(z_k) \subset A'$. Therefore we set $(v(y_k), v(z_k))$ to be any pair from $A_\alpha \times A_\beta$ or $A_\gamma \times A_\gamma$, respectively. Finally the second item of Lemma 2.16 allows us to extend v from V to V^*.

To see that such defined v is a solution of F^*, it remains to check that v satisfies the equations of the form $f_\alpha(x_i) \approx f_\alpha(x_j)$ created for $X_\alpha^i = X_\alpha^j$. However, the equivalence $(\star\star)$ yields $f_\alpha(v(x_i)) = f_\alpha(v(x_j))$, as required.

Conversely, take a solution $v : V^* \to A$ of the instance F^*. First of all, since v satisfies all equations in the sets $S_{A'}(x)$, we get $v(x) \in A'$ for all $x \in V$. To see that $s : X \to \{0, 1\}$ determined by:

$$s(X_\alpha^i) = \begin{cases} 1, & \text{if } v(x_i) \in A_\alpha, \\ 0, & \text{otherwise,} \end{cases}$$

is well defined suppose that $X_\alpha^i = X_\beta^j$.

- Case 1: $\alpha = \beta$. In this case the equation $f_\alpha(x_i) \approx f_\alpha(x_j)$ belongs to F^*. Since v is its solution, we get $f_\alpha(v(x_i)) = f_\alpha(v(x_j))$. Thus $v(x_i) \in A_\alpha \Leftrightarrow v(x_j) \in A_\alpha$, so that $s(X_\alpha^i) = s(X_\alpha^j)$, as required.
- Case 2: $\alpha \neq \beta$. We know that the set of equations $(\star)$ is satisfied by v. Since $v(x_i), v(x_j), v(y_k), v(z_k) \in A'$, one can check that every solution of this set is

forced to evaluate the pair $(f_\alpha(v(x_i)), f_\beta(v(x_j)))$ by either $(\top_\alpha, \top_\beta)$ or $(\bot_\alpha, \bot_\beta)$. Thus $v(x_i) \in A_\alpha \Leftrightarrow v(x_j) \in A_\beta$ so that $s(X_\alpha^i) = s(X_\beta^j)$, as required.

Observe that since $v(x_i) \in A'$, we have $v(x_i) \in A_\alpha$ for exactly one $\alpha \in \{1, 2, 3\}$. Thus $s(X_\alpha^i) = 1$ for exactly one $\alpha \in \{1, 2, 3\}$, which gives that every clause in F is satisfied according to POSITIVE- 1- IN- 3- SAT rules.   □

The next Lemma is a useful tool for characterizing unary algebras with SYS-TERMSAT in P. Actually it is a generalization of one of the reductions from the proof of Theorem 4.1. As there, we start here with an algebra $\mathbf{A} = (A, F)$ but define T to be the set of all generic operations from $Clo_1(\mathbf{A})$. We put $m = |T|$ so that T is listed by $(t_1, \ldots, t_m)$. First, for all $t \in T$ we choose and fix a function $\varphi_t : A \rightarrow \{0, 1\}$ satisfying:

• φ_t is bijective on $t(A)$,
• for all $f, g \in T$ if $f(A) = g(A)$ then $\varphi_f = \varphi_g$.

It is easy to see that such family of functions can always be chosen. Next we define the function $\varepsilon : A \rightarrow \{0, 1\}^m$ by $\varepsilon(a)_i = \varphi_{t_i}(t_i(a))$.

With the help of ε, for every $C \in \mathcal{C}(\mathbf{A})$ we define one m-ary relation R_C on $\{0, 1\}$ by putting:

$$R_C = \left\{\varepsilon(a) \in \{0, 1\}^m : a \in C\right\} = \left\{(\varphi_{t_1}(t_1(a)), \ldots, \varphi_{t_m}(t_m(a))) : a \in C\right\}.$$

Next we define a set $\mathcal{R}_0$ that consists of all relations of the form $R_C \subseteq \{0, 1\}^m$, where $C \in \mathcal{C}(\mathbf{A})$ and $|C| > 1$. Finally, similarly to the proof of Theorem 4.1, we put:

$$\mathcal{R} = \begin{cases} \mathcal{R}_0 \cup \{R_=\}, & \text{if the relational structure } (\{0, 1\}, \mathcal{R}_0) \text{ is proper} \\ & \text{or } \emptyset \notin \mathcal{C}(\mathbf{A}), \\ \{R_\bot\}, & \text{otherwise}, \end{cases}$$

where $R_= = \{(0, 0), (1, 1)\}$ and $R_\bot = \{(0, 1)\}$. The relational structure we need is defined by $\widehat{\mathbf{A}} = (\{0, 1\}, \mathcal{R})$.

Lemma 4.5 *Let* $\mathbf{A}$ *be a 2-valued unary algebra. Then the problem* SYSTERMSAT $(\mathbf{A})$ *can be polynomially transformed to* CSP $(\widehat{\mathbf{A}})$ *for the two-element relational structure* $\widehat{\mathbf{A}}$ *defined above.*

Proof To prove that the problem SYSTERMSAT $(\mathbf{A})$ can be polynomially transformed to CSP $(\widehat{\mathbf{A}})$, we start with an instance I of SYSTERMSAT $(\mathbf{A})$. At the beginning we apply Simplify procedure to I.

Suppose Simplify returns False, meaning that there is no solution of the instance I. We act analogously like in the proof of Theorem 4.1 and produce an unsatisfiable instance of CSP $(\widehat{\mathbf{A}})$. If Simplify returns True then we get an instance I' of CSYSTERMSAT $(\mathbf{A})$. Suppose I' has no equations so that we need to produce a satisfiable instace of CSP $(\widehat{\mathbf{A}})$. This time we also repeat the construction from the proof of Theorem 4.1.

After cleaning our setting from easy cases we are left with a nonempty I' that has all equations of the form $q(x) \approx r(y)$, where $q, r \in T$ (i.e. they are generic) and x, y are different variables. Moreover we have $|q(C(x)) \cap r(C(y))| > 1$. In particular $|q(A)| = |r(A)| = 2$ therefore $q(A) = r(A)$, and consequently $\varphi_q = \varphi_r$.

Let $V' := \mathrm{Var}(I') = \{x_1, \ldots, x_{v'}\}$. We construct an instance $I^\star$ of $\mathrm{CSP}(\widehat{A})$ consisting of:

- a set of variables $V = \left\{ X_j^i : x_i \in V' \text{ and } t_j \in T \right\}$,
- a set of constraints of two types:

  - $(R_=, (X_j^i, X_l^k))$, for each equation $t_j(x_i) \approx t_l(x_k)$ from I',
  - $(R_{C(x_i)}, (X_1^i, \ldots, X_m^i))$, for each variable $x_i \in V'$.

Since the constraint function C, and therefore the sets $C(x_i)$, are computed by the Simplify procedure, we know that $C(x_i) \in \mathcal{C}(A)$ and $|C(x_i)| > 1$ so that the relation $R_{C(x_i)} \in \mathcal{R}_0$.

To prove that $I^\star$ is equivalent to I' we start with a solution $v : V' \to A$ of I' and define a valuation $s : V \to \{0, 1\}$ by:

$$s(X_j^i) = \varphi_{t_j}(t_j(v(x_i))).$$

Consider a constraint of the form $(R_=, (X_j^i, X_l^k))$. Since v is a solution of I', the equation $t_j(x_i) \approx t_l(x_k)$ from I' is satisfied. Moreover $\varphi_{t_j} = \psi_{t_l}$, thus we get $s(X_j^i) = \varphi_{t_j}(t_j(v(x_i))) = \varphi_{t_l}(t_l(v(x_k))) = s(X_l^k)$, as required. Now consider a constraint of the form $(R_{C(x_i)}, (X_1^i, \ldots, X_m^i))$. Observe, that for all $t_j \in T$ we have $\varepsilon(v(x_i))_j = \varphi_{t_j}(t_j(v(x_i))) = s(X_j^i)$. We also have $v(x_i) \in C(x_i)$, so that $(s(X_1^i), \ldots, s(X_m^i)) = (\varepsilon(v(x_i))_1, \ldots, \varepsilon(v(x_i))_m) = \varepsilon(v(x_i)) \in R_{C(x_i)}$, as required.

Conversely, assume $s : V \to \{0, 1\}$ is a solution of $I^\star$ and pick a variable $x_i \in V'$. Since the constraint $(R_{C(x_i)}, (X_1^i \ldots, X_m^i))$ is satisfied by s we have $(s(X_1^i), \ldots, s(X_m^i)) \in R_{C(x_i)}$. By the definition of $R_{C(x_i)}$ there is $a_i \in C(x_i)$ such that $\varepsilon(a_i) = (s(X_1^i), \ldots, s(X_m^i))$. Define a valuation $v : V' \to A$ by putting $v(x_i) = a_i$. Thus $v(x_i) \in C(x_i)$ as required by the constraint function C. Now consider an equation $t_j(x_i) \approx t_l(x_k)$ from I', so that $\varphi_{t_j} = \varphi_{t_l}$. Since $(R_=, (X_j^i, X_l^k))$ is satisfied then $\varepsilon(v(x_i))_j = s(X_j^i) = s(X_l^k) = \varepsilon(v(x_k))_l$. It follows that $\varphi_{t_j}(t_j(v(x_i))) = \varepsilon(v(x_i))_j = \varepsilon(v(x_k))_l = \varphi_{t_l}(t_l(v(x_k)))$. Moreover $\varphi_{t_j}, \varphi_{t_l}$ are equal and bijective on $t_j(A) = t_l(A)$, thus $t_j(v(x_i)) = t_l(v(x_k))$, so that the equation $t_j(x_i) \approx t_l(x_k)$ is satisfied. This finishes the proof that each instance I of SYS-TERMSAT(A) is equivalent to some instance $I^\star$ of $\mathrm{CSP}(\widehat{A})$. $\qquad\square$

With the use of Lemma 4.5 sometimes we can classify the computational complexity of the problem SYSTERMSAT(A). This happens when, by Schaefer classification in Theorem 1.16, we get that $\mathrm{CSP}(\widehat{A})$ is in P. Then obviously SYSTERMSAT(A) is in P. Unfortunately, the situation when $\mathrm{CSP}(\widehat{A})$ is NP-complete gives us no information about the computational complexity of SYSTERMSAT(A). One way to overcome this situation is to make a 'good' choice for the functions of the form φ_t. Actually such a choice will be done in the proof of Theorem 4.7, when considering Subcase 8.3. However we do not know if such good φ_t's do exist in general, so that the corresponding $\widehat{\mathbf{A}}$ has $\text{CSP}(\widehat{\mathbf{A}})$ polynomially equivalent to $\text{SYSTERMSAT}(\mathbf{A})$. To ensure that they do exist, one needs a deeper look into the structure of 2-valued algebras. We have such understanding for $|A| = 4$. To get it, we need some preparations.

Suppose $R \subseteq \{0, 1\}^r$ and $1 \leqslant i < j \leqslant r$. We say that two coordinates i and j are R-dependent if either for all $a \in R$ we have $a_i = a_j$ or for all $a \in R$ we have $a_i \neq a_j$. In the first case we will say that the coordinates i and j are R-equal.

For $R \subseteq \{0, 1\}^r$ and the coordinate $1 \leqslant i \leqslant r$ we define $R^{(i)}$ to be the result of R after projection onto all but ith coordinate, i.e.:

$$R^{(i)} = \{(a_1, \ldots, a_{i-1}, a_{i+1}, \ldots, a_r) : (a_1, \ldots, a_{i-1}, a_i, a_{i+1}, \ldots, a_r) \in R$$
$$\text{for some } a_i \in \{0, 1\}\}.$$

We also denote by $a^{(i)} \in R^{(i)}$ the tuple $a \in R$ after this projection.

The following Lemma will be helpful in analysis of the computational complexity of a two-element CSP domain.

Lemma 4.6 *Let $R \subseteq \{0, 1\}^r$ (see Definition 1.14).*

(1) If $|R| \leqslant 2$ then R is affine and bijunctive.
(2) If $r \leqslant 2$ then R is bijunctive.
(3) If there are two R-dependent coordinates i, j then R is bijunctive (affine) if and only if $R^{(i)}$ or $R^{(j)}$ is bijunctive (affine).
(4) If there are two R-equal coordinates i, j then R is Horn (dual-Horn) if and only if $R^{(i)}$ or $R^{(j)}$ is Horn (dual-Horn).

Proof Recall from Fact 1.15 that the relation R is bijunctive, affine, Horn or dual-Horn if and only if R is preserved by the function $(x \vee y) \wedge (y \vee z) \wedge (x \vee z)$, $x \oplus y \oplus z$, $x \wedge y$ or $x \vee y$, respectively.

To prove (1) first take the function $x \oplus y \oplus z$. Observe that $|R| \leqslant 2$ implies that at least two arguments of the function $x \oplus y \oplus z$ are equal, so that it returns the remaining one. Therefore $x \oplus y \oplus z$ is a polymorphism of R and consequently R is affine. A similar argument shows that R is bijunctive.

Now observe that in (2) with $r \leqslant 2$ we have $|R| \leqslant 4$. In view of (1) we may additionally assume that $|R| > 2$. On the other hand, when $|R| = 4$ then R has all possible tuples, so that R is closed under every function, in particular R is bijunctive. Finally, suppose $|R| = 3$. We have got only four possibilities for the relation R. Moreover we only need to check whether the ternary majority function $(x \vee y) \wedge (y \vee z) \wedge (x \vee z)$, when applied to three different tuples $a, b, c \in R$ returns one of them. This can be done by checking all possible situations for a, b, c.

Now we are going to prove (3) and (4) for two R-equal coordinates i, j. As the arguments for bijunctive, affine and dual-Horn relations are pretty similar we present one only for Horn relations. First assume that R is Horn, so that it is preserved by $x \wedge y$. Now consider the relation $R^{(i)}$ and two tuples $a^{(i)}, b^{(i)} \in R^{(i)}$. Since the function $x \wedge y$ is applied component-wise we have:

$$a^{(i)} \wedge b^{(i)} = (a \wedge b)^{(i)}.$$

Thus $x \wedge y$ is a polymorphism of $R^{(i)}$, so that $R^{(i)}$ is Horn. Now suppose $R^{(i)}$ is Horn so that it admits $x \wedge y$ polymorphism. For convenience, without loss of generality we may assume that $i = 1$ and $j = 2$. To check whether R is Horn we take two tuples $a, b \in R$:

$$
\begin{aligned}
a \wedge b &= (a_1, a_2, \ldots, a_r) \wedge (b_1, b_2, \ldots, b_r) \\
&= (a_1 \wedge b_1, a_2 \wedge b_2, \ldots, a_r \wedge b_r) \\
&= (a_2 \wedge b_2, a_2 \wedge b_2, \ldots, a_r \wedge b_r).
\end{aligned}
$$

On the other hand, we know that $(a_2 \wedge b_2, \ldots, a_r \wedge b_r) \in R^{(1)}$. However by the definition of $R^{(1)}$ we get that for some $c_1 \in \{0, 1\}$ the tuple $(c_1, a_2 \wedge b_2, \ldots, a_r \wedge b_r) \in R$. Finally, since for every tuple c from R we have $c_1 = c_2$, we get $a_2 \wedge b_2 = c_1$ so that $a \wedge b \in R$ as required.

To prove (3) for R-dependent but not R-equal coordinates i, j we present an argument only for affine relations, as for bijunctive ones there is analogous one. Since i, j are R-dependent and not R-equal, we have $a_i \neq a_j$ for all $a \in R$, so that $a_i = a_j \oplus 1$. First assume that R is affine to argue that $R^{(i)}$ is affine by using $a^{(i)} \oplus b^{(i)} \oplus c^{(i)} = (a \oplus b \oplus c)^{(i)}$. Now suppose $R^{(i)}$ is affine so that it admits $x \oplus y \oplus z$ as a polymorphism. For convenience, without loss of generality we may assume that $i = 1$ and $j = 2$. To check whether R is affine we take three tuples $a, b, c \in R$:

$$
\begin{aligned}
a \oplus b \oplus c &= (a_1, a_2, \ldots, a_r) \oplus (b_1, b_2, \ldots, b_r) \oplus (c_1, c_2, \ldots, c_r) \\
&= (a_1 \oplus b_1 \oplus c_1, a_2 \oplus b_2 \oplus c_2, \ldots, a_r \oplus b_r \oplus c_r) \\
&= ((a_2 \oplus 1) \oplus (b_2 \oplus 1) \oplus (c_2 \oplus 1), a_2 \oplus b_2 \oplus c_2, \ldots, a_r \oplus b_r \oplus c_r) \\
&= (a_2 \oplus b_2 \oplus c_2 \oplus 1, a_2 \oplus b_2 \oplus c_2, \ldots, a_r \oplus b_r \oplus c_r).
\end{aligned}
$$

On the other hand, we know that $(a_2 \oplus b_2 \oplus c_2, \ldots, a_r \oplus b_r \oplus c_r) \in R^{(1)}$. However by the definition of $R^{(1)}$ we get that for some $d_1 \in \{0, 1\}$ the tuple $(d_1, a_2 \oplus b_2 \oplus c_2, \ldots, a_r \oplus b_r \oplus c_r) \in R$. Finally, since for every tuple d from R we have $d_1 = d_2 \oplus 1$, we get $a_2 \oplus b_2 \oplus c_2 \oplus 1 = d_1$ so that $a \oplus b \oplus c \in R$, as required. $\square$

Now we are ready to state the main result of this section that confirms Dichotomy Conjecture in the case of 2-valued four-element unary algebras.

Theorem 4.7 *Let* $\mathbf{A}$ *be a four-element 2-valued unary algebra. Then* SYSTERMSAT$(\mathbf{A}_A)$ *is either in* P *or is* NP-*complete.*

Proof Let T be the set of all generic operations from $Clo_1(\mathbf{A}_A)$, which is also the set of all generic operations from $Clo_1(\mathbf{A})$. Our complete characterization of computational complexity of SYSTERMSAT$(\mathbf{A}_A)$ for four-element 2-valued unary algebras is based on the family of kernels of operations from T, i.e. on the family $\mathcal{K}(\mathbf{A}) = \{Ker(t) : t \in T\}$.

Let $A = \{0, 1, 2, 3\}$. Since every $t \in T$ is 2-valued, the kernel of t consists of two equivalence classes. For example, if t is described by the following table:

x	t
0	α
1	α
2	β
3	β

then $Ker(t)$ is denoted by $01|23$. The order of classes and the order of elements inside a class are meaningless. Observe that in our setting every kernel is of the form $ab|cd$ or $abc|d$, to which we refer as $2 + 2$ or $3 + 1$ kernels, respectively.

The set $\mathcal{K}(\mathbf{A})$ contains at most 7 elements, that is four $3 + 1$ kernels and three $2 + 2$ kernels:

$$\mathcal{K}(\mathbf{A}) \subseteq \{123|0, 023|1, 013|2, 012|3, 01|23, 02|13, 03|12\}.$$

Our computational complexity characterization of the problem $\text{SYSTERMSAT}(\mathbf{A}_A)$ is based on the shape of $\mathcal{K}(\mathbf{A})$. In particular we are interested in the number of $3 + 1$ and $2 + 2$ kernels, which leads us to the following nine cases:

Case 1: $\mathcal{K}(\mathbf{A})$ contains all four $3 + 1$ kernels for term operations t_0, t_1, t_2 and t_3, with $Ker(t_i)$ leaving i as a singleton class.

In this case $\text{SYSTERMSAT}(\mathbf{A}_A)$ is NP-complete. We use Lemma 4.4 with $A_i = \{i\}$, $f_i = t_i$, $\top_i = t_i(i)$ and $\bot_i \neq \top_i$ (the other value of t_i). Observe that $A_1 \cup A_2 \cup A_3 = \{1, 2, 3\}$ is a definable constraint in the algebra $\mathbf{A}_A$ by the equation $t_0(x) \approx t_0(1)$.

Case 2: $\mathcal{K}(\mathbf{A})$ contains three $3 + 1$ kernels and at least one $2 + 2$ kernel.

In this case $\text{SYSTERMSAT}(\mathbf{A}_A)$ is NP-complete. Without loss of generality we may assume that $123|0, 023|1, 013|2, 03|12 \in \mathcal{K}(\mathbf{A})$ are kernels of t_0, t_1, t_2 and t_3, respectively. Again we use Lemma 4.4 with $A_i = \{i\}$ and $f_i = t_i$. As previously $A_1 \cup A_2 \cup A_3 = \{1, 2, 3\}$ is a constraint definable by the equation $t_0(x) \approx t_0(1)$.

Case 3: $\mathcal{K}(\mathbf{A})$ contains exactly three $3 + 1$ kernels and no $2 + 2$ kernel.

Without loss of generality we may assume that $123|0 \notin \mathcal{K}(\mathbf{A})$. Moreover, either the constraint $\{1, 2, 3\}$ is definable or not. In the first case we can repeat the argument of Case 1 to get NP-completeness of $\text{SYSTERMSAT}(\mathbf{A}_A)$. In the case when $\{1, 2, 3\} \notin \mathcal{C}(\mathbf{A}_A)$ we will show that $\text{SYSTERMSAT}(\mathbf{A}_A)$ is in P.

We use Lemma 4.5 to transform $\text{SYSTERMSAT}(\mathbf{A}_A)$ to $\text{CSP}(\widehat{\mathbf{A}_A})$, with $\widehat{\mathbf{A}_A} = (\{0, 1\}, \mathcal{R})$. We claim that $\mathcal{R}$ is bijunctive. Obviously $R_\bot$ and $R_=$ are bijunctive. Now consider a relation $R_C \in \mathcal{R}_0$. If $|C| = 2$ then R_C is bijunctive thanks to Lemma 4.6(1). Next suppose $|C| > 2$ and observe that coordinates determined by term operations from T with the same kernels are R_C-dependent. Therefore, when checking bijunctive property we can use Lemma 4.6(3) to switch to the appropriate $R_C^{(i)}$. For convenience let us denote by R_C' the relation R_C after eliminating R_C-dependent coordinates. For $a \in \{0, 1, 2, 3\}$ we denote by $\varepsilon'(a)$ the tuple $\varepsilon(a)$ after eliminating these R_C-dependent coordinates.

First suppose that $|C| = 4$, so that $C = A$. After eliminating R'_A-dependent coordinates we are left with three of them, corresponding to three different kernels. Next observe that Lemma 4.6(3) actually allows us to consider R'_A up to negation of any coordinate. Thus to check whether R_A is bijunctive we only need to check if the relation:

$$R'_A = \left\{\varepsilon'(0), \varepsilon'(1), \varepsilon'(2), \varepsilon'(3)\right\} = \{(0,0,0), (1,0,0), (0,1,0), (0,0,1)\}$$

is bijunctive. However R'_A admits the ternary majority polymorphism.

Now let $|C| = 3$ but $C \neq \{1, 2, 3\}$. Then, up to permuting coordinates, $R'_C = \{(0,0,0), (1,0,0), (0,1,0)\}$, so that R'_C is closed under majority function. As all relations in $\mathcal{R}$ are bijunctive we use Theorem 1.16 to conclude that $\mathrm{CSP}(\widehat{\mathbf{A}_A})$ is in P, so that, by Lemma 4.5, SYSTERMSAT$(\mathbf{A}_A)$ is in P, as well.

Case 4: $\mathcal{K}(\mathbf{A})$ contains two $3 + 1$ kernels and at least two $2 + 2$ kernels.

Without loss of generality we may assume that $123|0, 023|1 \in \mathcal{K}(\mathbf{A})$, are kernels of t_1 and t_2, respectively. We fall into two subcases, both of them NP-complete.

- Subcase 4.1: $01|23 \in \mathcal{K}(\mathbf{A})$ and this is witnessed by a term operation t_3. We use Lemma 4.4 with $A_1 = \{0\}$, $A_2 = \{1\}$, $A_3 = \{2, 3\}$ and $f_i = t_i$. This time $A_1 \cup A_2 \cup A_3 = A$, so that this is obviously a definable constraint.
- Subcase 4.2: $01|23 \notin \mathcal{K}(\mathbf{A})$, so that $02|13, 03|12 \in \mathcal{K}(\mathbf{A})$ and this is witnessed by t_0 and t_3, respectively. This time we use Lemma 4.4 with $A_i = \{i\}$, $f_1 = t_2$, $f_2 = t_0$ and $f_3 = t_3$. Observe that $A_1 \cup A_2 \cup A_3$ is a constraint definable by the equation $t_1(x) \approx t_1(1)$.

Case 5: $\mathcal{K}(\mathbf{A})$ contains exactly two $3 + 1$ kernels and exactly one $2 + 2$ kernel.

If all 3 kernels in $\mathcal{K}(\mathbf{A})$ are related as in Subcase 4.1 then SYSTERMSAT$(\mathbf{A}_A)$ is NP-complete. Thus we may assume, without loss of generality that $123|0, 023|1, 02|13 \in \mathcal{K}(\mathbf{A})$. In this case SYSTERMSAT$(\mathbf{A}_A)$ is in P. Indeed, we can repeat arguments from Case 3 to get that $\mathcal{R}$ consists of bijunctive relations.

Case 6: $\mathcal{K}(\mathbf{A})$ contains at most two kernels.

In this case SYSTERMSAT$(\mathbf{A}_A)$ is in P. Again, we can repeat arguments from Case 3. Indeed, after eliminating dependent coordinates we are left with at most two of them. Thus Lemma 4.6(2) applies and $\mathcal{R}$ consists of bijunctive relations.

Case 7: $\mathcal{K}(\mathbf{A})$ contains one $3 + 1$ kernel and three $2 + 2$ kernels.

In this case SYSTERMSAT$(\mathbf{A}_A)$ is NP-complete. Without loss of generality we may assume that $123|0, 01|23, 02|13, 03|12 \in \mathcal{K}(\mathbf{A})$ and this is witnessed by t_0, t_1, t_2 and t_3, respectively. We again use Lemma 4.4 with $A_i = \{i\}$ and $f_i = t_i$. As previously $A_1 \cup A_2 \cup A_3$ is a definable constraint by the equation $t_0(x) \approx t_0(1)$.

Case 8: $\mathcal{K}(\mathbf{A})$ contains exactly one $3 + 1$ kernel and exactly two $2 + 2$ kernels.

Without loss of generality we may assume that the kernels $012|3, 01|23, 02|13$ are in $\mathcal{K}(\mathbf{A})$ and this is witnessed by t_3, t_1 and t_2, respectively. First suppose that $\{1, 2, 3\}$ is definable. Then we apply Lemma 4.4 with $A_i = \{i\}$ and $f_i = t_i$ to conclude that SYSTERMSAT$(\mathbf{A}_A)$ is NP-complete.

Otherwise, if $\{1, 2, 3\}$ is not definable, we need to look closer into the structure of the operations from T. All we know is that the kernels are determined by the following behavior of t_1, t_2 and t_3:

x	t_1	t_2	t_3
0	$\top_1$	$\top_2$	$\top_3$
1	$\top_1$	$\bot_2$	$\top_3$
2	$\bot_1$	$\top_2$	$\top_3$
3	$\bot_1$	$\bot_2$	$\bot_3$

Subcase 8.1: The set $\{1, 2\}$ is definable.

In this situation SYSTERMSAT$(\mathbf{A}_A)$ is NP-complete. We present a reduction from 3- SAT into SYSTERMSAT$(\mathbf{A}_A)$. Take a formula $F = C_1 \wedge \cdots \wedge C_n$, where $C_i = L_1^i \vee L_2^i \vee L_3^i$ and L_α^i is a variable or a negated variable. We denote by X the set of variables of F. We start the construction of an instance $F^\star$ of SYSTERMSAT$(\mathbf{A}_A)$ by creating variables $x_1^i, x_2^i, x_3^i, u_1^i, u_2^i, u_3^i$ for $i \in \{1, \ldots, n\}$. Next, for each clause C_i, we add the following set of equations to $F^\star$, which will help to provide proper clause valuation:

$$Q_i = \begin{cases} t_1(x_1^i) \approx t_1(u_1^i) \\ t_2(x_2^i) \approx t_2(u_1^i) \\ t_2(x_3^i) \approx t_2(u_3^i) \\ t_3(u_1^i) \approx t_3(u_2^i) \\ t_1(u_3^i) \approx t_1(u_2^i) \\ t_3(u_3^i) \approx \top_3. \end{cases}$$

To code dependencies between 3- SAT literals we first define sets of equations $E_{(\alpha,\beta)}(x, y)$ and $N_{(\alpha,\beta)}(x, y)$, where x, y are two variables and α, β range over $\{1, 2, 3\}$. The variables of equations in $E_{(\alpha,\beta)}(x, y)$ are among x, y, z, w, while those of $N_{(\alpha,\beta)}(x, y)$ are among x, y, z, x', y', where z, w, x', y' are completely new variables and are chosen to be different in each set of the form $E_{(\alpha,\beta)}(x, y)$ or $N_{(\alpha,\beta)}(x, y)$.

- $E_{(\alpha,\beta)}(x, y) = E_{(\beta,\alpha)}(y, x)$, for $\alpha, \beta \in \{1, 2, 3\}$,
- $E_{(\alpha,\alpha)}(x, y) = \{t_\alpha(x) \approx t_\alpha(y)\}$, for $\alpha \in \{1, 2, 3\}$,
- for $\alpha \in \{1, 2\}$ we put:

$$E_{(\alpha,3)}(x, y) = \begin{cases} t_\alpha(z) \approx t_\alpha(x) \\ t_{3-\alpha}(z) \approx \bot_{3-\alpha} \\ t_3(z) \approx t_3(y), \end{cases}$$

- $E_{(1,2)}(x, y) = E_{(1,3)}(x, w) \cup E_{(3,2)}(w, y)$.

Before defining $N_{(\alpha,\beta)}(x, y)$ recall that Lemma 2.16 supplies us with the set $S_{\{1,2\}}(z)$ of equations that forces value of z to be in the definable constraint $\{1, 2\}$. Now for $\alpha, \beta \in \{1, 2, 3\}$ we put:

$$N_{(\alpha,\beta)}(x, y) = E_{(\alpha,1)}(x, x') \cup S_{\{1,2\}}(z) \cup E_{(2,\beta)}(y', y)$$
$$\cup \{t_1(x') \approx t_1(z), t_2(y') \approx t_2(z)\}.$$

Claim 4.8 *A valuation v of variables x and y can be extended to a solution of the set of equations $E_{(\alpha,\beta)}(x, y)$ if and only if the following equivalence holds:*

$$t_\alpha(v(x)) = \mathsf{T}_\alpha \Leftrightarrow t_\beta(v(y)) = \mathsf{T}_\beta.$$

Analogously, a valuation v of variables x and y can be extended to a solution of the set of equations $N_{(\alpha,\beta)}(x, y)$ if and only if the following equivalence holds:

$$t_\alpha(v(x)) = \mathsf{T}_\alpha \Leftrightarrow t_\beta(v(y)) = \perp_\beta.$$

Proof The first part of the Claim can be easily checked by considering all possibilities. The second part, for $N_{(\alpha,\beta)}(x, y)$, is only a bit harder. Since $S_{\{1,2\}}(z)$ forces $v(z) \in \{1, 2\}$, we get:

$$t_1(v(x')) = \mathsf{T}_1 \Leftrightarrow t_2(v(y')) = \perp_2.$$

Finally, we use the first part of our Claim to get:

$$t_\alpha(v(x)) = \mathsf{T}_\alpha \Leftrightarrow t_1(v(x')) = \mathsf{T}_1 \Leftrightarrow t_2(v(y')) = \perp_2 \Leftrightarrow t_\beta(v(y)) = \perp_\beta,$$

as required. □

Now we are ready to impose the dependencies on variables x_1^i, x_2^i, x_3^i whenever $L_\alpha^i = L_\beta^j$ or $L_\alpha^i = \neg L_\beta^j$. For $L_\alpha^i = L_\beta^j$ we add to $F^\star$ the set $E_{(\alpha,\beta)}(x_\alpha^i, x_\beta^j)$. If $L_\alpha^i = \neg L_\beta^j$ then we add the set $N_{(\alpha,\beta)}(x_\alpha^i, x_\beta^j)$ to $F^\star$. Now, when all such dependencies are covered and $F^\star$ is completely constructed, we denote $\mathrm{Var}(F^\star)$ by $V^\star$.

To prove that F and $F^\star$ are equivalent first take a solution $s : X \to \{0, 1\}$ of F to define a valuation $v : V^\star \to A$. We put:

$$v(x_\alpha^i) = \begin{cases} 0, & \text{if } s(L_\alpha^i) = 1, \\ 3, & \text{otherwise.} \end{cases}$$

to get the following equivalence:

$$t_\alpha(v(x_\alpha^i)) = \mathsf{T}_\alpha \Leftrightarrow s(L_\alpha^i) = 1.$$

Since s satisfies the clause C_i we get that v valuates at least one of the variables x_1^i, x_2^i, x_3^i to 0. Thanks to this observation one can check, by examining Q_i, that we can extend v to the variables u_1^i, u_2^i, u_3^i since they occur only in Q_i.

Now suppose $L_\alpha^i = L_\beta^j$ so that F^* contains the set $E_{(\alpha,\beta)}(x_\alpha^i, x_\beta^j)$. Since $s(L_\alpha^i) = s(L_\beta^j)$, we have:

$$t_\alpha(v(x_\alpha^i)) = \top_\alpha \Leftrightarrow s(L_\alpha^i) = 1 \Leftrightarrow s(L_\beta^j) = 1 \Leftrightarrow t_\beta(v(x_\beta^j)) = \top_\beta$$

and we use Claim 4.8 to extend v to a solution of the set $E_{(\alpha,\beta)}(x_\alpha^i, x_\beta^j)$.

If $L_\alpha^i = \neg L_\beta^j$ then F^* contains the set $N_{(\alpha,\beta)}(x_\alpha^i, x_\beta^j)$. Since $s(L_\alpha^i) = 1 - s(L_\beta^j)$, we have:

$$t_\alpha(v(x_\alpha^i)) = \top_\alpha \Leftrightarrow s(L_\alpha^i) = 1 \Leftrightarrow s(L_\beta^j) = 0 \Leftrightarrow t_\beta(v(x_\beta^j)) = \bot_\beta$$

and again we use Claim 4.8 to extend v to a solution of the set $N_{(\alpha,\beta)}(x_\alpha^i, x_\beta^j)$.

Now take a solution $v : V^* \to A$ of F^* to define a valuation $s : X \to \{0, 1\}$ of F. We put $s(L_\alpha^i) = 1$ if and only if $t_\alpha(v(x_\alpha^i)) = \top_\alpha$. To observe that s is well defined, i.e. the value of L_α^i is determined correctly if $L_\alpha^i \in \left\{ L_\beta^j, \neg L_\beta^j \right\}$ we simply invoke Claim 4.8. It remains to show that s satisfies all clauses C_i. Assume to the contrary that this is not the case for some i, i.e. $s(L_1^i) = s(L_2^i) = s(L_3^i) = 0$. Thus $t_\alpha(v(x_\alpha^i)) = \bot_\alpha$, for all $\alpha \in \{1, 2, 3\}$. But then the set Q_i has no solution, contrary to the fact that v is one of its solutions.

We have just finished the proof that F and F^* are equivalent, so that SYS-TERMSAT$(\mathbf{A}_A)$ is NP-complete.

Subcase 8.2: The set $\{1, 2\}$ is not definable and there are $f, g \in T$ such that $f(A) = g(A)$ and $f(0) \neq g(0)$.

This time SYSTERMSAT$(\mathbf{A}_A)$ is NP-complete by a reduction going also from 3-SAT into SYSTERMSAT$(\mathbf{A}_A)$. We follow the proof of Subcase 8.1. Definability of $\{1, 2\}$ was used only in the definition of the $N_{(\alpha,\beta)}(x, y)$, when we invoked the set $S_{\{1,2\}}(z)$. To overcome this, first note that the $E_{(\alpha,\beta)}(x, y)$ remains unchanged and satisfy first part of Claim 4.8. Moreover $Ker(f)$, $Ker(g)$ are in $\mathcal{K}(\mathbf{A})$ and therefore, for some $\gamma, \delta \in \{1, 2, 3\}$, they are just $Ker(t_\gamma)$ and $Ker(t_\delta)$, respectively. Looking at the kernels in $\mathcal{K}(\mathbf{A})$ we see that $f(0) \neq g(0)$ gives $f(0) = g(3)$. We now put:

$$N_{(\alpha,\beta)}(x, y) = E_{(\alpha,\gamma)}(x, x') \cup \left\{ f(x') \approx g(y') \right\} \cup E_{(\delta,\beta)}(y', y).$$

For every solution v of $N_{(\alpha,\beta)}(x, y)$ we have:

$$\begin{aligned} t_\alpha(v(x)) = \top_\alpha &\Leftrightarrow t_\gamma(v(x')) = \top_\gamma \\ &\Leftrightarrow f(v(x')) = f(0) \\ &\Leftrightarrow g(v(y')) = f(0) \end{aligned}$$

$$\Leftrightarrow g(v(y')) = g(3)$$
$$\Leftrightarrow t_\delta(v(y')) = \perp_\delta$$
$$\Leftrightarrow t_\beta(v(y)) = \perp_\beta,$$

so that Claim 4.8 holds for redefined N. Equipped with $E_{(\alpha,\beta)}(x, y)$'s and $N_{(\alpha,\beta)}$ (x, y)'s satisfying Claim 4.8 the rest of the proof that 3- SAT can be interpreted in SYSTERMSAT $(\mathbf{A}_A)$ remains unchanged.

Subcase 8.3: The set $\{1, 2\}$ is not definable and for all $f, g \in T$ such that $f(A) = g(A)$ we have $f(0) = g(0)$.

In this situation SYSTERMSAT $(\mathbf{A}_A)$ is in P. We are going to use Lemma 4.5, but this time when constructing $\mathcal{R}_0$ we choose the functions φ_t with much more care. On the top of the two conditions they have to satisfy, we want to have $\varphi_t(t(0)) = 0$, for all $t \in T$. Thanks to the second condition defining Subcase 8.3 this can be easily done.

We claim that $\mathcal{R}$ constructed with these φ_t's is Horn. The relations $R_\perp$ and $R_=$ are obviously Horn. Now take $R_C \in \mathcal{R}_0$ and observe that coordinates determined by term operations from T with the same kernels are R_C-equal. Indeed, thanks to our condition $\varphi_t(t(0)) = 0$, such coordinates are not only R_C-dependent (as in Case 3) but in fact R_C-equal. Similarly to Case 3, when checking Horn property for R_C, we use Lemma 4.6(4) to switch to $R_C^{(i)}$ and finally to R'_C and $\varepsilon'(a)$ as defined there.

Now to prove that $\mathcal{R}_0$ is Horn, we are left with the following relation:

$$R'_A = \left\{\varepsilon'(0), \varepsilon'(1), \varepsilon'(2), \varepsilon'(3)\right\} = \{(0, 0, 0), (0, 1, 0), (1, 0, 0), (1, 1, 1)\}$$

and some of its subsets R_C if $|C| < 4$. Since R'_A is closed under $x \wedge y$ operation, it is Horn. Next consider C such that $2 \leqslant |C| \leqslant 3$. Since $C \neq \{1, 2\}$ and $C \neq \{1, 2, 3\}$, we know that $\{1, 2\} \subseteq C$ implies $0 \in C$. Thus R'_C is closed under $x \wedge y$ and therefore R'_C is Horn. Finally CSP $(\widehat{\mathbf{A}_A})$ is in P, so that SYSTERMSAT $(\mathbf{A}_A) \in$ P, as well.

Case 9: $\mathcal{K}(\mathbf{A})$ contains no $3 + 1$ kernel and three $2 + 2$ kernels.

If some three-element constraint is definable then we can repeat our argument from Case 7 to get NP-completeness of SYSTERMSAT $(\mathbf{A}_A)$. Otherwise we show that SYSTERMSAT $(\mathbf{A}_A)$ is in P by repeating arguments from Case 3. Observe that now we have $|C| = 2$ or $|C| = 4$. However, this time we argue that $\widehat{\mathbf{A}_A}$ is affine instead of bijunctive.

This concludes our computational complexity characterization of SYSTERMSAT $(\mathbf{A}_A)$ for 2-valued four-element unary algebras. $\qquad\square$

Theorem 4.7 is formulated as to confirm the Dichotomy Conjecture. However, from the proof we can actually get the classification described in the following table:

$3+1$ \ $2+2$	3	2	1	0
4	Case 1	Case 1	Case 1	Case 1
3	Case 2	Case 2	Case 2	Case 3
2	Case 4	Case 4	Case 5	Case 6
1	Case 7	Case 8	Case 6	Case 6
0	Case 9	Case 6	Case 6	Case 6

The rows and columns correspond to the number of $3 + 1$ and $2 + 2$ kernels, respectively. The white background of the cells in the above table stands for the polynomial complexity, dark background for NP-complete cases and gray one, on the diagonal, for the mixed cases (details can be found in the proof).

Theorems 4.2 and 4.7 cover SYSPOLSAT for a wide class of four-element unary algebras. Unfortunately SYSPOLSAT(**A**) for **A** that is 3-valued, but is neither strongly-3-valued nor 2-valued remains open. This seems to be the hardest part of the four-element unary algebras classification. The next picture graphically describes situations, in which the computational complexity of SYSPOLSAT for four-element unary algebras is known. First of the six circular diagrams represents all possible 256 unary term operations over four-element set. The circle is divided into several regions, which represent operations with the same image, e.g. $\{1, 2, 3\}$ in the smallest central circular region. There is one exception for this rule, i.e. the innermost ring contains all constants and permutations. The regions are chosen in such a way, that their areas are proportional to the number of corresponding operations. The number of constants and permutations is obviously $4 + 24 = 28$, while the number of operations that take exactly two fixed values is 14. Finally, there are 36 operations that have a fixed three-element image.

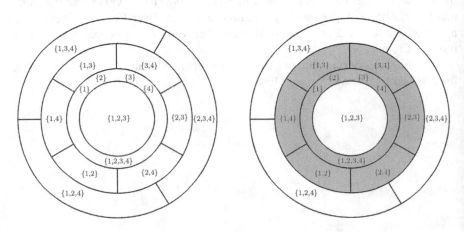

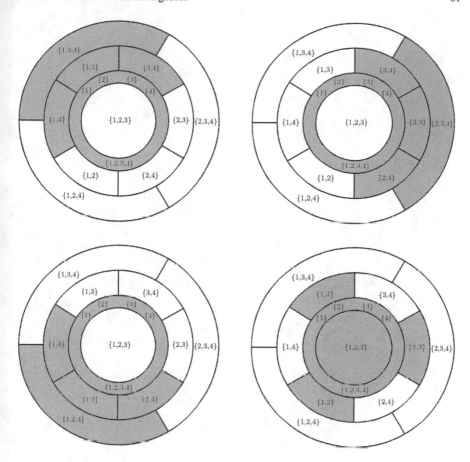

The second diagram presents the result of Theorem 4.7. Indeed, if all term operations from the clone of a four-element unary algebra **A** belong to the gray regions then we can classify the computational complexity of the problem SYSPOLSAT(**A**). The same applies to the next four diagrams, which present the known situations that come from Theorem 4.2.

References

[Bul06] Bulatov AA (2006) A dichotomy theorem for constraint satisfaction problems on a 3-element set. J ACM 53(1):66–120 (electronic)

[Sch78] Schaefer TJ (1978) The complexity of satisfiability problems. In: Conference record of the tenth annual ACM symposium on theory of computing, ACM, San Diego, New York, pp 216–226

Chapter 5
Conclusions and Open Problems

Abstract We present conclusions and open problems raising from studying solving equations over unary algebras. We suggest areas that are most promising for expanding our knowledge.

We have not solved the Meta-Problem for SYSTERMSAT over unary algebras nor even get close to such a solution. However we have found two important equivalences:

- The characterization of computational complexity of solving equations over unary algebras is as hard as the classification of CONSTRAINT SATISFACTION PROBLEM over all relational structures.
- The characterization of unary algebras with polynomial time SYSTERMSAT is as hard as of all algebras.

On the other hand the tools we have worked out allowed us to confirm Dichotomy Conjecture for solving equations in many special cases. In particular we have shown:

- Several characterization theorems for unary algebras, which term operations take only few values.
- A transparent characterization of computational complexity for three-element unary algebras.
- A classification of a broad class of four-element unary algebras.

We hope that our work will help to push further the knowledge about solving equations over unary algebras. In particular the following open problems seem to be the most vulnerable to attack:

1. A classification of arbitrary large 2-valued algebras. We think that this task is a starting point for a new attack.
2. A classification of all four-element algebras. We guess that a deep exploration of Bulatov's classification for CSP should be very helpful, although technical.

Finally, we present two main open problems in the area:

© The Author(s) 2015
P. Broniek, *Computational Complexity of Solving Equation Systems*,
SpringerBriefs in Philosophy, DOI 10.1007/978-3-319-21750-5_5

3. A full classification of solving equations over unary algebras. If that shows up to be too hard, every new partial characterization might also be interesting.
4. What is the computational complexity of Meta-Problem for solving equations (over unary algebras)?

Printed in the United States
By Bookmasters